IVOR GURNEY
Selected Poems

Also available:

Collected Poems of Ivor Gurney (ed. P. J. Kavanagh, 1982; revised, 1984)
Michael Hurd, *The Ordeal of Ivor Gurney* (1978; reissued 1990)

Selected Poems of
IVOR GURNEY

Selected and introduced by
P. J. KAVANAGH

Oxford   New York
OXFORD UNIVERSITY PRESS
1990

Oxford University Press, Walton Street, Oxford OX2 6DP

Oxford New York Toronto
Delhi Bombay Calcutta Madras Karachi
Petaling Jaya Singapore Hong Kong Tokyo
Nairobi Dar es Salaam Cape Town
Melbourne Auckland

and associated companies in
Berlin Ibadan

Oxford is a trade mark of Oxford University Press

First issued as an Oxford University Press paperback 1990

British Library Cataloguing in Publication Data
Gurney, Ivor, 1890–1937
Selected poems of Ivor Gurney – (Oxford Poems)
I. Title   II. Kavanagh, P. J. (Patrick Joseph, 1931–   )
821′ .912
ISBN 0–19–282636–0

Library of Congress Cataloging-in-Publication Data
Gurney, Ivor, 1890–1937
[Poems. Selections. 1990]
Selected poems of Ivor Gurney / selected and introduced by P. J. Kavanagh
p.   cm.
I. Kavanagh. P. J. (Patrick Joseph), 1931–   . II. Title
PR6013.U693A6 1990   821′.912—dc19   89–22831
ISBN 0–19–282636–0

Typeset by Wyvern Typesetting Ltd, Bristol
Printed in Great Britain by
J. W. Arrowsmith Ltd., Bristol

NOTE ON SELECTION

This selection, of 153 poems, is made from the *Collected Poems of Ivor Gurney*, ed. P. J. Kavanagh (Oxford University Press, 1982). The text is that of the revised paperback edition (1984), with some further revisions.

CONTENTS

INTRODUCTION

THE first selection of Ivor Gurney's poetry was published in 1954, long enough ago to have decided what it is like, how universal its appeal. But nearly a quarter of a century later he was still being described, when a plaque to him was unveiled in Gloucester Cathedral, as a 'local' poet. It was well-meant—he was a local boy— but in the limiting sense of Edward Thomas's definition of poets 'whom we can connect with a district of England and often cannot sunder from it without harm', Gurney is not a local poet at all.

Nevertheless, like most poets, he is dependent on the particular and on being able to name it. After the First World War, when he could name the places in France where he had served, his war poetry leaps to life. He had found wartime censorship more than usually cramping: 'forbidden names or dates without which the poets / Are done for . . .'. His awareness that everything is happening in one place rather than another, at a certain hour, under a never-to-be-repeated pattern of sky, are what give these poems vitality. Whereas the other war poets (Owen, Sassoon, and so on) reacted against the war rhetoric of their elders with indignation and tell us truths we ought to have guessed, Gurney gives us pictures we would not have imagined: the gentleness of his first reception in the front line, the effect of a clarinet played in the trenches. It is the poetry of a particularized, not a generalized, humanity, of the flesh and nerves rather than of the intellect. His precisions can be journalistic, but he almost invariably looks up to notice the behaviour of the French sky. This widening of view after such narrowness of observation can be startling, putting the war itself in its place.

Why, then, did he not at once take his place among his fellow-poets of the war, and why has it taken so long for his work to gain acceptance? Apart from his own disinclination to compete, or pretend, several other things were working against him. He was, for instance, acknowledged to be a musician, a composer of songs, of genius—and we always doubt whether someone can be good at two things. He was also mentally unbalanced: from about 1912, when he was twenty-two, he had been subject to breakdowns, and in 1922 he was put in an asylum—where he wrote some of his best poems—for the last fifteen years of his life, until his death in 1937. So, Gurney was not only 'primarily' a musician (and a 'local' poet and an odd sort of war poet), he was a 'mad' poet too, and the combination of all these things caused him to be shunted off into a siding; at least for a time.

Most of the poems selected here show no sign of mental

disturbance at all—unless to be a poet is to be in such a condition, which is possible, and certainly Gurney showed himself unsuited to the routines of 'ordinary' life. Where there are signs of unbalance, they are obvious; too many of his preoccupations crowd in at once and fall over each other: his sense of having been betrayed, his memories of his comrades in France, of Gloucestershire, of walks at night. The result, though seldom completely incoherent, is confused and painful. But these are also the themes of his best work and, right up to the end of his working life, he is capable of sudden, pictorial, simplicities.

Ivor Bertie Gurney (Bertie was a family surname) was born at 3 Queen Street, Gloucester, on 28 August 1890, the son of a tailor. A young clergyman, Alfred Cheesman, stood as godparent at his christening, and this was to have a significant effect on Gurney's life. As he grew, Cheesman took him under his protection, lending him books, walking and talking with him, and became the first of the surrogate parents Gurney was to acquire. So much so that he became a mystery to his own family; in the words of his sister: 'The truth was, he did not seem to belong to us. . . . He simply called on us briefly, and left again without a word!'

He was educated, as a chorister, at King's School, Gloucester, and in 1911 won an open scholarship to the Royal College of Music in London. The impression he made at the College was of erratic brilliance. His biographer, Michael Hurd, quotes Sir Charles Stanford: 'In later years Sir Charles declared that of all his pupils—Vaughan Williams, Ireland, Bliss and dozens more—Gurney was potentially "the biggest of them all. But [he added] he was the least teachable."' In 1912 Gurney began to set poems to music, including five Elizabethan lyrics. Of these Hurd, himself a musician, says: 'Gurney jumped in one bound from mere competence to mastery and genuine originality.' He adds that in his view the inspiration does not seem to be a musical one at all, but in direct response to the innocence and freedom of the poetry. So it is not surprising that about this time Gurney started to write verses himself. His Gloucestershire friends, John Haines and F. W. Harvey, already wrote; John Masefield had recently had a great popular success with *The Everlasting Mercy*: poetry was on one of its periodic upward swings.

At this time too came the first signs that the pendulum of his own moods could swing too violently. 'The Young Genius does not feel too well,' he wrote to Mariot Scott, a music historian, editor of the RCM magazine, and his mentor, 'and his brain won't move as he wishes it to.' He fled from London, to Gloucestershire and his writing friends, where he talked, walked, wrote, smoked, drank tea, starved himself (he was always poor) and then compulsively ate cream buns. In contrast, with the Chapman family (whom he met in 1914 when he took an organist's post at High Wycombe), he was a cheerful, playful member of a very normal group.

Then the Great War began. Gurney volunteered, was first turned down because of his eyesight, and then in 1915 was accepted. Many of his letters of the time are high-spirited and often funny; but others suggest how bad the previous period had also been: 'It is a better way to die; with these men, in such a cause; than the end which seemed near me and was so desirable only just over two years ago.' That may be the self-dramatization of a talented young man; but it is enough to dispose of the idea that Gurney's subsequent mental illness was entirely owing to the war.

He served as a Private in the 2nd/5th Gloucestershire Regiment, and his war is very much that of the private soldier: a cleaner business, humanly speaking, than that of someone with more responsibility. A conscripted Private's job in war is, rightly, to obey orders and stay alive, meanwhile making himself as comfortable as possible. In other words, to remain (unlike the Officer candidates in 'The Bohemians') triumphantly a civilian.

He sent back poems and song-settings; and with Marion Scott's help his first book, *Severn and Somme*, was published in 1917. In the same year he was mildly wounded, mildly gassed; spent time in hospitals, and was discharged unfit shortly before the armistice.

The post-war years, from 1918 to 1922, are a graph of mental disturbance and recovery. He went back to the Royal College for a time, to study under Vaughan Williams, and resumed his organist's post, and friendship with the Chapmans, at High Wycombe; but he was too restless, and took to wandering, occasionally sleeping on the Embankment or walking through the night back to Gloucester (he was justifiably proud of being a 'night-walker'). Early in this period his second volume of poems, *War's Embers*, was published, but he remained desperately short of cash. He had a small pension (twelve shillings a week) and friends helped, but they could not help enough. Like many men returned from France—and here the war surely does play its part—there is something inconsolable about him at this time. He returned to Gloucester and worked as a cinema pianist, farm labourer, tax clerk; but nothing lasted for long. Sometimes he settled: with his aunt at Longford outside Gloucester, then in a farm cottage on the slopes of the Cotswolds under his beloved Crickley Hill ('Felling a Tree' gives some idea of his life there and the hopes he had of it). But he set himself an unbalanced regime; he seems to have decided, for instance, to go without sleep.

Friends rallied to him: Gloucestershire ones, influential London ones—Vaughan Williams, Walter de la Mare, J. C. Squire, and others—but nothing seemed to suffice. Possibly this is because he was too preoccupied. Sleepless or not, during this time—1919 to 1922—he was doing some of his best work in verse (he also wrote much music): 'Drachms and Scruples', 'Walking Song', 'Cotswold Ways', 'Water Colours', 'Between the Boughs', 'The Silent One', 'The Lock Keeper', 'Longford Dawns', 'Time to Come', 'Yesterday Lost', 'The Hoe Scrapes Earth', 'The High Hills', 'Clay'—all belong to this period. So, as often happens in the story of Gurney, just when we begin to see pathos in his situation, and perhaps unconsciously to patronise him, he satisfactorily eludes us.

Worse than pathos was to come. In September 1922 he was committed to Barnwood House in Gloucester, a private asylum for the insane. In December he was moved to the City of London Mental Hospital at Dartford, Kent. He never saw Gloucestershire again. But he continued to write. The next three or four years contain some of his most original poems. He becomes, for instance, the master of first lines. To go through them is like reading one huge mysterious poem. No poet could live up to so many splendid beginnings. Grandiloquent ones, such as 'Darkness has cheating swiftness', 'What evil coil of fate has fastened me', 'Smudgy dawn scarfed with military colours', and interestingly conversational ones, 'One comes across the strangest things in walks'. He is also capable of magnificent, poem-saving, last lines. In between, there are sometimes flaws, which are obvious: flat phrases, quirky ones, confusing syntax. These are sometimes unfortunate later additions, but not always. Even at his most contorted, his general meaning usually comes clear if we persist, carried along by the energy of the poem, not stopping, to the end; and individual obscurities clarify themselves when we go back. Gurney however did write some perfect poems (like 'The Songs I Had'). He is good at the short sprint, the poem expressed in two or three breaths.

But perfection of that kind is not what he was interested in. He is not a Georgian poet who 'broke down' but one who consciously, though unprogrammatically, broke away and was, as far as he knew, on his own, fortified by his beloved Whitman. His subjects are conventional, but the intensity with which he sees and expresses them is not. He knew this, and one can sense his impatience with his more popular contemporaries because their matter—hedgerows, skies, 'Nature'—he considered peculiarly his own. So, 'free of useless fashions', he wishes to go behind their verses, behind the verse of preceding centuries, back to his masters, the Elizabethans and Jacobeans, as the Lake Poets went back to the Ballads. The impatience is in his language, as though he wished to be as free as the Elizabethans in fashioning a new one. He wants a poetry composed on the nerves, 'a book that brings the clear / Spirit of him that wrote', and he hurls himself headlong, so that we feel 'the football rush of him', as he said in admiration of George Chapman. The result is, that although Gurney tells us surprisingly little about himself, we feel he has entirely opened his heart; there is nothing prudently withheld, and he is always, unpredictably, musical. Late poems, such as 'Traffic in Sheets' and 'Sea-Marge', are almost pure music.

If we are, as we think we are, an age that admires honesty, Gurney's time has come. There is seldom any striving for effect, any 'putting it on' (his phrase), in Gurney; none at all after 1920. He wrung the neck of his early elegance ('no swank'). What may have seemed naive and unpolished to his contemporaries has for us the stamp of sincerity. Technically, musically, he is more in command than he would wish us, at first, to notice. Like Whitman, he was concerned with finding and touching the core of innocence in his own nature, and he addresses himself to ours.

CHRONOLOGY

1890 28 August: Ivor Bertie born in Gloucester. Second of four children of David Gurney (tailor) from Maisemore on the River Severn, and Florence Lugg, from Bisley in the Cotswold hills.

1900–06 In Gloucester Cathedral Choir and King's School.

1906–11 Articled pupil to Cathedral organist. Tour of English cathedrals with his godfather and patron, Revd Alfred Cheesman. Wins scholarship to Royal College of Music, London.

1911–14 At RCM. Compositions include settings of Elizabethan songs. Meets Marion Scott (1877–1953), music historian. Part-time job as church organist at High Wycombe, and friendship with Chapman family there.

August 1914: War declared. Gurney volunteers (at first unsuccessfully).

1914–18 February 1915: Drafted into 2nd/5th Gloucester Regiment. With battalion in Northampton, Chelmsford etc. Plays bass cornet (baryton) in military band.

February 1916: Training on Salisbury Plain. May 1916: Gloucesters cross to Le Havre, march to Front Line. Service in various parts of Northern France.

Writes many letters; poems and music sent to Marion Scott who acts as his agent. Some songs performed; poems accepted for publication by Sidgwick & Jackson.

April 1917: Gurney wounded in arm; to hospital in Rouen.

September 1917: in gas near Passchendaele; sent home ('Blighty'); in hospital near Edinburgh; meets Annie Drummond, a nurse.

October 1917: *Severn and Somme* (46 poems) published, to good reviews.

Training courses in north of England, with spells in hospital.

June 1918: Breakdown; suicide letters. October 1918: Discharged with 'deferred shell-shock'. Returns to Gloucester. Small Army pension.

November 1918: War ends.

1919–22 May 1919: Gurney's father dies. *War's Embers* (58 poems) published ('To His Love' chiefly praised; some complaints about colloquialisms). New collection of poems rejected. Returns to RCM; lodgings at Earl's Court; later in High Wycombe. Visits John Masefield with F. W. Harvey.

Most productive years of music composition and of poems. Songs published. Takes various unsuccessful jobs in London and Gloucestershire (cinema pianist, in cold storage depot, tax clerk, on farm). Stays with aunt at Longford, near Gloucester; later moves in with his brother. Increasing signs of mental disturbance.

September 1922: Committed to asylum at Gloucester.

December 1922: Moved to City of London Mental Hospital at Dartford, Kent.

1922–37 In mental hospital. Reading, writing of poems, verse-autobiography, and 'Appeals' [for his release], music.

Marion Scott ensures all possible MSS preserved. Song-cycles published; a few poems in magazines.

After 1926, fewer extant MSS; latest date 1933.

Music and Letters magazine celebrating Gurney printed just before his death on 26 December 1937.

31 December 1937: Ivor Gurney buried at Twigworth, near Gloucester.

1938–90 Publication and radio programmes of his music gradually proceed. Gerald and Joy Finzi undertake promotion of his work and cataloguing of music and poems.

78 poems, ed. Edmund Blunden (Hutchinson), 1954.

MSS donated to Gloucester Library by Gurney's brother, Ronald, 1959.

140 poems ed. Leonard Clark (Chatto & Windus), 1973.

The Ordeal of Ivor Gurney, biography by Michael Hurd (OUP), 1978; reissued paperback, 1990.

Collected Poems, ed. P. J. Kavanagh (OUP), 1982; revised 1984.

War Letters, ed. R. K. R. Thornton (Carcanet), 1983.

Letters to the Chapman Family, ed. Anthony Boden (Alan Sutton), 1986.

11 November 1985: Ivor Gurney's name on memorial to Poets of the First World War unveiled in Westminster Abbey.

Severn and Somme and *War's Embers*, ed. R. K. R. Thornton, reissued (Carcanet), 1987.

Selected Poems (the present edition), ed. P. J. Kavanagh (OUP), 1990, to mark centenary of Gurney's birth.

Selected Poems

Bach and the Sentry

Watching the dark my spirit rose in flood
  On that most dearest Prelude of my delight.
The low-lying mist lifted its hood,
  The October stars showed nobly in clear night.

When I return, and to real music-making,
  And play that Prelude, how will it happen then?
Shall I feel as I felt, a sentry hardly waking,
  With a dull sense of No Man's Land again?

Song

    Only the wanderer
      Knows England's graces,
    Or can anew see clear
      Familiar faces.

    And who loves joy as he
      That dwells in shadows?
    Do not forget me quite,
      O Severn meadows.

3

Ballad of the Three Spectres

As I went up by Ovillers
   In mud and water cold to the knee,
There went three jeering, fleering spectres,
   That walked abreast and talked of me.

The first said, 'Here's a right brave soldier
   That walks the dark unfearingly;
Soon he'll come back on a fine stretcher,
   And laughing for a nice Blighty.'

The second, 'Read his face, old comrade,
   No kind of lucky chance I see;
One day he'll freeze in mud to the marrow,
   Then look his last on Picardie.'

Though bitter the word of these first twain
   Curses the third spat venomously;
'He'll stay untouched till the war's last dawning
   Then live one hour of agony.'

Liars the first two were. Behold me
   At sloping arms by one—two—three—
Waiting the time I shall discover
   Whether the third spake verity.

Pain

Pain, pain continual; pain unending;
Hard even to the roughest, but to those
Hungry for beauty . . . Not the wisest knows,
Nor most pitiful-hearted, what the wending
Of one hour's way meant. Grey monotony lending
Weight to the grey skies, grey mud where goes
An army of grey bedrenched scarecrows in rows
Careless at last of cruellest Fate-sending.
Seeing the pitiful eyes of men foredone,
Or horses shot, too tired merely to stir,
Dying in shell-holes both, slain by the mud.
Men broken, shrieking even to hear a gun.
Till pain grinds down, or lethargy numbs her,
The amazed heart cries angrily out on God.

Servitude

If it were not for England, who would bear
This heavy servitude one moment more?
To keep a brothel, sweep and wash the floor
Of filthiest hovels were noble to compare
With this brass-cleaning life. Now here, now there
Harried in foolishness, scanned curiously o'er
By fools made brazen by conceit, and store
Of antique witticisms thin and bare.

Only the love of comrades sweetens all,
Whose laughing spirit will not be outdone.
As night-watching men wait for the sun
To hearten them, so wait I on such boys
As neither brass nor Hell-fire may appal,
Nor guns, nor sergeant-major's bluster and noise.

To His Love

He's gone, and all our plans
   Are useless indeed.
We'll walk no more on Cotswold
   Where the sheep feed
   Quietly and take no heed.

His body that was so quick
   Is not as you
Knew it, on Severn river
   Under the blue
   Driving our small boat through.

You would not know him now . . .
   But still he died
Nobly, so cover him over
   With violets of pride
   Purple from Severn side.

Cover him, cover him soon!
   And with thick-set
Masses of memoried flowers—
   Hide that red wet
   Thing I must somehow forget.

Turmut-Hoeing

I straightened my back from turmut-hoeing
  And saw, with suddenly opened eyes,
Tall trees, a meadow ripe for mowing,
  And azure June's cloud-circled skies.

Below, the earth was beautiful
  Of touch and colour, fair each weed,
But Heaven's high beauty held me still,
  Only of music had I need.

And the white-clad girl at the old farm,
  Who smiled and looked across at me,
Dumb was held by that strong charm
  Of cloud-ships sailing a foamless sea.

Ypres—Minsterworth

(To F.W.H.)

Thick lie in Gloucester orchards now
   Apples the Severn wind
With rough play tore from the tossing
   Branches, and left behind
Leaves strewn on pastures, blown in hedges,
   And by the roadway lined.

And I lie leagues on leagues afar
   To think how that wind made
Great shoutings in the wide chimney,
   A noise of cannonade—
Of how the proud elms by the signpost
   The tempest's will obeyed—

To think how in some German prison
   A boy lies with whom
I might have taken joy full-hearted
   Hearing the great boom
Of autumn, watching the fire, talking
   Of books in the half gloom.

O wind of Ypres and of Severn
   Riot there also, and tell
Of comrades safe returned, home-keeping
   Music and autumn smell.
Comfort blow him and friendly greeting,
   Hearten him, wish him well!

Old Martinmas Eve

The moon, one tree, one star.
Still meadows far,
Enwreathed and scarfed by phantom lines of white.
November's night
Of all her nights, I thought, and turned to see
Again that moon and star-supporting tree.
If some most quiet tune had spoken then;
Some silver thread of sounds; a core within
That sea-deep silentness, I had not known
Even such joy in peace, but sound was none—
Nor should be till birds roused to find the dawn.

He talked of Africa,
    That fat and easy man.
I'd but to say a word,
    And straight the tales began.

And when I'd wish to read,
    That man would not disclose
A thought of harm, but sleep;
    Hard-breathing through his nose.

Then when I'd wish to hear
    More tales of Africa,
'Twas but to wake him up,
    And but a word to say

To press the button, and
    Keep quiet; nothing more;
For tales of stretching veldt,
    Kaffir and sullen Boer.

O what a lovely friend!
    O quiet easy life!
I wonder if his sister
    Would care to be my wife. . . .

The Battalion Is Now On Rest

(To 'La Comtesse')

Walking the village street, to watch the stars and find
Some peace like the old peace, some soothe for soul and mind;
The noise of laughter strikes me as I move on my way
Towards England—westward—and the last glow of day.

And here is the end of houses. I turn on my heel,
And stay where those voices a moment made me feel
As I were on Cotswold, with nothing else to do
Than stare at the old houses, to taste the night-dew;

To answer friendly greetings from rough voices kind . . .
Oh, one may try for ever to be calm and resigned,
A red blind at evening sets the poor heart on fire—
Or a child's face, a sunset—with the old hot desire.

Above Ashleworth

O does some blind fool now stand on my hill
To see how Ashleworth nestles by the river?
Where eyes and heart and soul may drink their fill.

The Cotswolds stand out eastward as if never
A curve of them the hand of Time might change,
Beauty sleeps most confidently for ever.

The blind fools stands, his dull eyes free to range
Endlessly almost, and finds no word to say:
Not that the sense of wonder is too strange

Too great for speech. Naught touches him; the day
Blows its glad trumpets, breathes rich-odoured breath
Glory after glory passes away

(And I'm in France!). He looks, and sees beneath
The clouds in steady Severn silver and grey
But dead he is, and comfortable in death.

O *Tree of Pride*

O Tree of pride,
Before your green to gold and orange fade,
And scarce one single leaf of summer's shade
Remains to hide
Robin or wren,
Give me one song of all your songs, that men
May take your beauty winter's fire beside.

For memory passes
Of even the loveliest things, bravest in show;
The mind to beauty most alert not know
How the August grasses
Waved, by December's
Glow, unless he see deep in the embers
The poet's dream, gathered from cold print's spaces.

The Companions

On uplands bleak and bare to wind
Beneath a maze of stars one strode.
Phantoms of fear haunted the road,
Mocking my footsteps close behind.

Till Heaven blew clear of cloud, showed each
Most tiny baby-star as fine
As any king's jewel, Orion
Triumphed through tracery of beech.

So unafraid the tramp went on
Past dusky rut and pool alight,
With Heaven's chief wonder of night,
Jupiter close companion.

And in no mood of pride, courteous,
Light-hearted as with a king's friend
He went with me till journey's end
His courtiers Mars and Regulus.

My door reached, gladly had I paid
With stammered thanks his courtesy,
And theirs, but ne'er a star could see
Of all Heaven's ordered cavalcade.
Pools inky-black, unbottomed shade . . .
Fine snow drove west and blinded me.

Michaelmas

The autumn rooms of green and bronze
Are swept with cleanest airs today.
Large air puffs jostle little ones—
Not quarrelsome nor yet in play.

And in the valley bonfires spread
A blue enchantment on the day.
No spoil, no flaw! How good that spring
Is lengths of calendar away.

When from the Curve

When from the curve of the wood's edge does grow
Power, and that spreads to envelope me—
Wrapped up in sense of meeting tree and plough
I feel tiny song stir tremblingly
And deep; the many birth-pangs separate
Taking most full of joy, for soon shall come
The kindling, the beating at Heaven gate
The flood of tide that bears strongly home.

Then under the skies I make my vows
Myself to purify and fit my heart
For the inhabiting of the high House
Of Song, that dwells high and clean apart.
The fire, the flood, the soaring, these the three
That merged are power of song and prophesy.

Advice

Why do you not steer straight, my love, she cried:
The wind makes steady your way, favours the tide:
The boat obeys the helm, were you now to steer
Courageous, our troubles and doubts would vanish here.

And cassia and pearls would pack your hold,
And your returning act crown manifold
Our upward course, and not a thing to desire . . .
The rudder swang in the tide, and we beached in the mire.

First Time In

After the dread tales and red yarns of the Line
Anything might have come to us; but the divine
Afterglow brought us up to a Welsh colony
Hiding in sandbag ditches, whispering consolatory
Soft foreign things. Then we were taken in
To low huts candle-lit, shaded close by slitten
Oilsheets, and there the boys gave us kind welcome,
So that we looked out as from the edge of home.
Sang us Welsh things, and changed all former notions
To human hopeful things. And the next day's guns
Nor any line-pangs ever quite could blot out
That strangely beautiful entry to war's rout;
Candles they gave us, precious and shared over-rations—
Ulysses found little more in his wanderings without doubt.
'David of the White Rock', the 'Slumber Song' so soft, and that
Beautiful tune to which roguish words by Welsh pit boys
Are sung—but never more beautiful than there under the guns'
    noise.

La Gorgue

The long night, the short sleep, and La Gorgue to wander,
So be the Fates were kind and our Commander;
With a mill, and still canal, and like-Stroudway bridges.
One looks back on these as Time's truest riches
Which were so short an escape, so perilous a joy
Since fatigues, weather, line trouble or any whimsical ploy
Division might hatch out would have finished peace.

There was a house there, (I tell the noted thing)
The kindest woman kept, and an unending string
Of privates as wasps to sugar went in and out.
Friendliness sanctified all there without doubt,
As lovely as the mill above the still green
Canal where the dark fishes went almost unseen.
B Company had come down from Tilleloy
Lousy, thirsty, avid of any employ
Of peace; and this woman in leanest times had plotted
A miracle to amaze the army-witted.
And this was café-au-lait as princes know it,
And fasting, and poor-struck; dead but not to show it.
A drink of edicts, dooms, a height of tales.
Heat, cream, coffee; the maker tries and fails,
The poet too, where such thirst such mate had.
A campaign thing that makes remembrance sad.

There was light there, too, in the clear North French way.
It blessed the room, and bread, and the mistress giver,
The husband for his wife's sake, and both for a day
Were blessed by many soldiers tired however;
A mark in Time, a Peace, a Making-delay.

Songs Come to the Mind

Songs come to the mind—
Other men's songs
Or one's own, when something is kind
And remembers not any wrongs.

Swift cleaving paths in air
On a bicycle, or slow
Wandering and wondering where
One's purposes may go.

Songs come and are taken, written,
Snatched from the momentary
Accidents of light, shape, spirit meeting
For one light second spirit, unbelievably.

That Centre of Old

Is it only Cotswold that holds the glamour
Memory felt of England in the gun-stammer—
Thud, smack, belch of war—and kept virtue by?
I do not know, but only that, most unhappy,
The hills are to me what to happy I
They were in Somme muckage-baths and east of Laventie
When hunger made one worthy to absorb the sky-
Look, or play fancy-tricks with small cloudlets high:
Count them—or dare not count—love and let go by.
Now as ever Cotswold rewards the mere being and seeing
As truly as
Ever in the relief of knowing mere being
In the still space
At a strafe end grateful for silence and body's grace
(Whole body—and after hell's hammering and clamouring).
Then memory purified made rewarding shapes
Of all that spirit runs towards in escapes,
And Cooper's Hill showed plain almost as experience.
Soft winter mornings of kind innocence, high June's
Girl's air of untouched purity, and on Cooper's Hill
Or autumn Cranham with its boom of colour . . .
Not anyway does ever Cotswold's fail—her dear blue long
        dark slope fail—
Of the imagining promise in full exile.

Possessions

France has victory, England yet firm shall stay,
But what shall please the wind now the trees are away
War took on Witcombe steep?
It breathes there, and wonders at old roarings
October time at all lights; and the new clearings
For memory are like to weep.
War need not cut down trees, three hundred miles over seas,
Children of those the Romans saw—lovely trunk and great-
    sail trees!
Not on Cranham, not on Cooper's of camps;
Friend to the great October stars—and the July sky lamps.

Billet

O, but the racked clear tired strained frames we had!
Tumbling in the new billet on to straw bed,
Dead asleep in eye shutting. Waking as sudden
To a golden and azure roof, a golden ratcheted
Lovely web of blue seen and blue shut, and cobwebs and tiles,
And grey wood dusty with time. June's girlish kindest smiles.
Rest at last and no danger for another week, a seven-day week.
But one Private took on himself a Company's heart to speak,
'I wish to bloody hell I was just going to Brewery—surely
To work all day (in Stroud) and be free at tea-time—allowed
Resting when one wanted, and a joke in season,
To change clothes and take a girl to Horsepool's turning,
Or drink a pint at "Travellers Rest", and find no cloud.
Then God and man and war and Gloucestershire would have
      a reason,
But I get no good in France, getting killed, cleaning off mud.'
He spoke the heart of all of us—the hidden thought burning,
      unturning.

The Soaking

The rain has come, and the earth must be very glad
Of its moisture, and the made roads all dust clad;
It lets a friendly veil down on the lucent dark,
And not of any bright ground thing shows any spark.

Tomorrow's grey morning will show cow-parsley,
Hung all with shining drops, and the river will be
Duller because of the all soddenness of things,
Till the skylark breaks his reluctance, hangs shaking, and sings.

First March

It was first marching, hardly we had settled yet
To think of England, or escaped body pain—
Flat country going leaves but small chance for
The mind to escape to any resort but its vain
Own circling greyness and stain.
First halt, second halt, and then to spoiled country again.
There were unknown kilometres to march, one must settle
To play chess or talk home-talk or think as might happen.
After three weeks of February frost few were in fettle,
Barely frostbite the most of us had escapen.
To move, then to go onward, at least to be moved.
Myself had revived and then dulled down, it was I
Who stared for body-ease at the grey sky
And watched in grind of pain the monotony
Of grit, road metal, slide underneath by.
To get there being the one way not to die.
Suddenly a road's turn brought the sweet unexpected
Balm. Snowdrops bloomed in a ruined garden neglected:
Roman the road as of Birdlip we were on the verge,
And this west country thing so from chaos to emerge.
One gracious touch the whole wilderness corrected.

Laventie

One would remember still
Meadows and low hill
Laventie was, as to the line and elm row
Growing through green strength wounded, as home elms grow.
Shimmer of summer there and blue autumn mists
Seen from trench-ditch winding in mazy twists.
The Australian gunners in close flowery hiding
Cunning found out at last, and smashed in the unspeakable lists.
And the guns in the smashed wood thumping and grinding.
The letters written there, and received there,
Books, cakes, cigarettes in a parish of famine,
And leaks in rainy times with general all-damning.
The crater, and carrying of gas cylinders on two sticks
(Pain past comparison and far past right agony gone),
Strained hopelessly of heart and frame at first fix.

Café-au-lait in dug-outs on Tommies' cookers,
Cursed minniewerfs, thirst in eighteen-hour summer.
The Australian miners clayed, and the being afraid
Before strafes, sultry August dusk time than death dumber—
And the cooler hush after the strafe, and the long night wait—
The relief of first dawn, the crawling out to look at it,
Wonder divine of dawn, man hesitating before Heaven's gate.
(Though not on Cooper's where music fire took at it.
Though not as at Framilode beauty where body did shake at it)
Yet the dawn with aeroplanes crawling high at Heaven gate
Lovely aerial beetles of wonderful scintillate
Strangest interest, and puffs of soft purest white—
Seeking light, dispersing colouring for fancy's delight.
Of Machonachie, Paxton, Tickler and Gloucester's Stephens;
Fray Bentos, Spiller and Baker, odds and evens
Of trench food, but the everlasting clean craving
For bread, the pure thing, blessèd beyond saving.
Canteen disappointments, and the keen boy braving
Bullets or such for grouse roused surprisingly through
(Halfway) Stand-to.
And the shell nearly blunted my razor at shaving;
Tilleloy, Fauquissart, Neuve Chapelle, and mud like glue.
But Laventie, most of all, I think is to soldiers
The town itself with plane trees, and small-spa air;

And vin, rouge-blanc, chocolat, citron, grenadine:
One might buy in small delectable cafés there.
The broken church, and vegetable fields bare;
Neat French market-town look so clean,
And the clarity, amiability of North French air.

Like water flowing beneath the dark plough and high Heaven,
Music's delight to please the poet pack-marching there.

The Touchstone—Watching Malvern

What Malvern is the day is, and its touchstone—
Grey velvet, or moon-marked; rich, or bare as bone;
One looks towards Malvern and is made one with the whole;
The world swings round him as the Bear to the Pole.

Men have crossed seas to know how Paul's tops Fleet,
That as music has rapt them in the mere street,
While none or few care how the curved giants stand,
(Those upheaved strengths!) on the meadow and plough-land.

Glimmering Dusk

Glimmering dusk above the moist plough and the
Silence of trees' heaviness under low grey sky
Are some comfort for the mind gone soft in lethargy;
White road dark with pools, but growing soon to dry;
But the mind complains 'This indeed is beauty enough
And comfort, but itself not enough cure for sorrows,
Nor equal weight for good things and fine stuff
Of thought snatched ruthlessly by thieves best under harrows'.
But the soul would not be denied; comfort from the night
Gathered, the mind unwilling, hope past all thought of
    matters
Of right anger—and the body only spoke of its plight
That a kind law makes dust of at last and scatters.
No notice the soul took—it desired God with all friendly might.
The leaves, not sodden, moved on trees with winter patters.

New Year's Eve

Aveluy and New Year's Eve, and the time as tender
As if green buds grew. In the low west a slender
Streak of last orange. Guns mostly deadest still.
And a noise of limbers near coming down the hill.
Nothing doing, nothing doing, and a screed to write,
Candles enough for books, a sleepy delight
In the warm dug-out, day ended. Nine hours to the light.
There now and then now, one nestled down snug.
A head is enough to read by, and cover up with a rug.
Electric. Clarinet sang of 'A Hundred Pipers'
And hush awe mystery vanished like tapers
Of tobacco smoke, there was great hilarity then!
Breath, and a queer tube, magicked sorrow from men.
The North, and all Scott called me—Ballads and Burns again!
Enough! I got up and lit (the last little bit
But one) of candle and poked the remaining fire,
Got some blaze into the cold; sat, wrote verse there . . .
(Or music). The 'hundred pipers' had called so plain
('And a'') and for three hours stuck it and worked as best
Drippings, and cold, and misery would let desire.

Darkness Has Cheating Swiftness

Darkness has cheating swiftness
When the eyes rove,
Opens and shuts in long avenues
That thought cannot prove.

Darkness shuts in and closes;
There are three ghosts
Different in one clump of hedge roses
And a threat in posts.

Until one tops the road crest,
Turns, sees the city lie
Long stretched out in bright sparkles of gratefullest
Homecalling array.

Daily—Old Tale

If one's heart is broken twenty times a day,
What easier thing than to fling the bits away,
But still one gathers fragments, and looks for wire,
Or patches it up like some old bicycle tyre.

Bicycle tyres fare hardly on roads, but the heart
Has an easier time than rubber, they sheathe a cart
With iron; so lumbering and slow my mind must be made,
To bother the heart and to teach things and learn it its trade.

After War

One got peace of heart at last, the dark march over,
And the straps slipped, the warmth felt under roof's low cover,
Lying slack the body, let sink in straw giving;
And some sweetness, a great sweetness felt in mere living.
And to come to this haven after sorefooted weeks,
The dark barn roof, and the glows and the wedges and streaks;
Letters from home, dry warmth and still sure rest taken
Sweet to the chilled frame, nerves soothed were so sore shaken.

Brimscombe

One lucky hour in middle of my tiredness
I came under the pines of the sheer steep
And saw the stars like steady candles gleam
Above and through them; Brimscombe wrapped (past life)
    in sleep:
Such body weariness and bad ugliness
Had gone before, such tiredness to come on me;
This perfect moment had such pure clemency
That it my memory has all coloured since,
Forgetting the blackness and pain so driven hence.
And the naked uplands from even bramble free,
That ringed-in hour of pines, stars and dark eminence.
Wonder of men had walked there, and old Romance.
(The thing we looked for in our fear of France.)

Riez Bailleul

Riez Bailleul in blue tea-time
Called back the Severn lanes, the roads
Where the small ash leaves lie, and floods
Of hawthorn leaves turned with night's rime.
No Severn though, nor great valley clouds ...

Now, in the thought, comparisons
Go with those here-and-theres and fancy
Sees on the china firelight dancy,
The wall lit where the sofa runs.
A dear light like Sirius or spring sun's.

But the trench thoughts will not go, tomorrow
Up to the line, and no straw laid
Soft for the body, and long nights' dread,
Lightless, all common human sorrow.
Unploughed the grown field once was furrow.

Meanwhile soft azure, and fall's emerald
Lovely the road makes, a softness clings
Of colour and texture of light; there rings
Metal, as it were, in air; and the called
Of twilight, dim stars of the dome, appear.

There's dusk here; west hedgerows show thin;
In billets there's sound of packs reset,
Tea finished; the dixies dried of the wet.
Some walk, some write, and the cards begin.
Stars gather in heaven and the pools drown in.

Canadians

We marched, and saw a company of Canadians,
Their coats weighed eighty pounds at least, we saw them
Faces infinitely grimed in, with almost dead hands
Bent, slouching downwards to billets comfortless and dim.
Cave dwellers last of tribes they seemed, and a pity
Even from us just relieved, much as they were, left us.
Lord, what a land of desolation, what iniquity
Of mere being, of what youth that country bereft us;
Plagues of evil lay in Death's Valley, we also
Had forded that up to the thighs in chill mud,
Gone for five days then any sign of life glow,
As the notched stumps or the grey clouds we stood
Dead past death from first hour and the needed mood
Of level pain shifting continually to and fro.
Saskatchewan, Ontario, Jack London ran in
My own mind; what in others? these men who finely
Perhaps had chosen danger for reckless and fine,
Fate had sent for suffering and dwelling obscenely
Vermin-eaten, fed beastly, in vile ditches meanly.

Drachms and Scruples

Misery weighed by drachms and scruples
Is but scrawls on a vain page.
To cruel masters are we pupils,
Escape comes careless with old age.

O why were stars so set in Heaven
To desire greedily as gluttons do,
Or children trinkets—May death make even
So rough an evil as we go through.

April Gale

The wind frightens my dog, but I bathe in it,
Sound, rush, scent of the spring fields.

My dog's hairs are blown like feathers askew,
My coat's a demon, torturing like life.

Compensations

Spring larch should set the body shaking
In masterless pleasure,
But virtue lies in a square making—
The making pleasure.
True, the poet's true place is in that high wood,
And his gaze on it,
But work has a bent, and some grey sort of good.
Worship, or a sonnet?

April—Dull Afternoon

The sun for all his pride dims out and dies.
Afternoon sees not one
Of all those flames that lit the primrose lamps
At winter's hest fordone,

Like music eager curving or narrowing
From here to there. Strange how no mist can dull
Wholly the silver edge of April song
Though the air's a blanket weighing on like wool.

Silver Birch

A silver birch dances at my window.
The faint clouds dimly seen
On the sloped azure are easy to be scattered
When full day's wind sweeps clean.

Call to walk comes as of true nature,
Easy should the body move.
And poetry comes after eight miles' seeking,
Mere right out of mere love.

The Garden

The ordered curly and plain cabbages
Are all set out like school-children in rows;
In six short weeks shall these no longer please,
For with that ink-proud lady the rose, pleasure goes.

I cannot think what moved the poet men
So to write panegyrics of that foolish
Simpleton—while wild rose as fresh again
Lives, and the drowsed cabbages keep soil coolish.

The Awakening

In the white painted dark lobby
The rosy firelight is thrown,
And the mat is still moisted with fresh mud
As I work at my task alone.

The murmuring of the kettle soothes me—
As those above sleep on still.
I love that dear winter-reflection . . .
Gone truant from loving too well.

Walking Song

The miles go sliding by
Under my steady feet,
That mark a leisurely
And still unbroken beat,
Through coppices that hear
Awhile, then lie as still
As though no traveller
Ever had climbed their hill.

My comrades are the small
Or dumb or singing birds,
Squirrels, field-things all
And placid drowsing herds.
Companions that I must
Greet for a while, then leave
Scattering the forward dust
From dawn to late of eve.

Cotswold Ways

One comes across the strangest things in walks:
Fragments of Abbey tithe-barns fixed in modern
And Dutch-sort houses where the water baulks
Weired up, and brick kilns broken among fern,
Old troughs, great stone cisterns bishops might have blessed
Ceremonially, and worthy mounting-stones;
Black timber in red brick, queerly placed
Where Hill stone was looked for—and a manor's bones
Spied in the frame of some wisteria'd house
And mill-falls and sedge pools and Saxon faces;
Stream-sources happened upon in unlikely places,
And Roman-looking hills of small degree
And the surprise of dignity of poplars
At a road end, or the white Cotswold scars
Or sheets spread white against the hazel tree.
Strange the large difference of up-Cotswold ways;
Birdlip climbs bold and treeless to a bend,
Portway to dim wood-lengths without end,
And Crickley goes to cliffs are the crown of days.

Quiet Talk

Tree-talk is breathing quietly today
Of coming autumn and the staleness over—
Pause of high summer when the year's at stay,
And the wind's sick that now moves like a lover.

On valley ridges where our beeches cluster
Or changing ashes guarding slopes of plough,
He goes now sure of heart, now with a fluster
Of teasing purpose. Night shall find him grow

To dark strength and a cruel spoiling will.
First he will baffle streams and dull their bright,
Cower and threaten both about the hill—
Before their death trees have their full delight.

Water Colours

The trembling water glimpsed through dark tangle
Of late-month April's delicatest thorn,
One moment put the cuckoo-flower to scorn
Where its head hangs by sedges, Severn bank-full.
But dark water has a hundred fires on it;
As the sky changes it changes and ranges through
Sky colours and thorn colours, and more would do,
Were not the blossom truth so quick on it,
And beauty brief in action as first dew.

Generations

The ploughed field and the fallow field
They sang a prudent song to me:
We bide all year and take our yield
Or barrenness as case may be.

What time or tide may bring to pass
Is nothing of our reckoning,
Power was before our making was
That had in brooding thought its spring.

We bide our fate as best betides
What ends the tale may prove the first.
Stars know as truly of their guides
As we the truth of best or worst.

Going Out at Dawn

Strange to see that usual dark road paven wet
With shallow dim reflecting rain pools looking
To north, where light all night stayed and dawn braving yet
Capella hung, above dark elms unshaking, no silence breaking.
And still to dawn night's ugliness owed no debt.
About eleven from the touch of the drear raining,
I had gone in to Shakespeare and my own writing,
Seen the lovely lamplight in golden shining,
And resolved to move no more till dawn made whitening
Between the shutter-chinks, or by the door-mat.
Yet here at five, an hour before day was alive . . .
Behold me walking to where great elm trees drip
Melancholy slow streams of rainwater, on the too wet
Traveller, to pass them, watching, and then return.
Writing Sonata or Quartett with a candle dip.

Possessions

Sand has the ants, clay ferny weeds for play
But what shall please the wind now the trees are away
War took on Witcombe steep?
It breathes there, and wonders at old night roarings;
October time at all lights, and the new clearings
For memory are like to weep.
It was right for the beeches to stand over Witcombe reaches,
Until the wind roared and softened and died to sleep.

When I Am Covered

When I am covered with the dust of peace
And but the rain to moist my senseless clay,
Will there be one regret left in that ill ease

One sentimental fib of light and day—
A grief for hillside and the beaten trees?
Better to leave them, utterly to go away.

When every tiny pang of love is counterpiece
To shadowed woe of huge weight and the stay
For yet another torment ere release

Better to lie and be forgotten aye.
In death his rose-leaves never is a crease.
Rest squares reckonings love set awry.

The Change

Gone bare the fields now, and the starlings gather,
Whirr above stubble and soft changing hedges.
Changed the season chord too, F major or minor,
The gnats sing thin in clouds above the sedges.

And there is nothing proud now, not disconsolate,
Nothing youthful save where dark crocus flings
Summer's last challenge toward winter's merciless
Cohort, for whom the robin alone sings.

Fields for a while longer, then, O soul,
A curtained room close shut against the rime—
Where shall float music, voice or violin's
Denial passionate of the frozen time.

Stars Sliding

The stars are sliding wanton through trees,
The sky is sliding steady over all.
Great Bear to Gemini will lose his place
And Cygnus over world's brink slip and fall.

Follow-my-Leader's not so bad a game.
But were it leap-frog: O to see the shoots
And tracks of glory; Scorpions and Swans tame
And Argo swarmed with Bulls and other brutes.

Of Grandcourt

Through miles of mud we travelled, and by sick valleys—
The Valley of Death at last—most evil alleys,
To Grandcourt trenches reserve—and the hell's name it did
    deserve.
Rain there was—tired and weak I was, glad for an end.
But one spoke to me—one I liked well as friend—
'Let's volunteer for the Front Line—many others won't.
I'll volunteer, it's better being there than here.'
But I had seen too many ditches and stood too long
Feeling my feet freeze, and my shoulders ache with the
    strong
Pull of equipment—and too much use of pain and strain.
Besides he was Lance Corporal and might be full Corporal
Before the next straw resting might come again,
Before the next billet should hum with talk and song.
Stars looked as well from second as from first line holes.
There were fatigues for change, and a thought less danger—
But five or six there were followed Army with their souls—
Took five days' dripping rain without let or finish again—
With dysentery and bodies of heroic ghouls.
Till at last their hearts feared nothing of the brazen anger,
(Perhaps of death little) but once more again to drop on straw
    bed-serving,
And to have heaven of dry feeling after the damps and fouls.

Between the Boughs

Between the boughs the stars showed numberless
And the leaves were
As wonderful in blackness as those brightnesses
Hung in high air.

Two lovers in that whispering silence, what
Should fright our peace?
The aloofness, the dread of starry majesties,
The night-stilled trees.

The Silent One

Who died on the wires, and hung there, one of two—
Who for his hours of life had chattered through
Infinite lovely chatter of Bucks accent;
Yet faced unbroken wires; stepped over, and went,
A noble fool, faithful to his stripes—and ended.
But I weak, hungry, and willing only for the chance
Of line—to fight in the line, lay down under unbroken
Wires, and saw the flashes, and kept unshaken.
Till the politest voice—a finicking accent, said:
'Do you think you might crawl through, there; there's a hole?' In
   the afraid
Darkness, shot at; I smiled, as politely replied—
'I'm afraid not, Sir.' There was no hole, no way to be seen.
Nothing but chance of death, after tearing of clothes.
Kept flat, and watched the darkness, hearing bullets whizzing—
And thought of music—and swore deep heart's deep oaths
(Polite to God—) and retreated and came on again.
Again retreated—and a second time faced the screen.

The Lock Keeper

Men delight to praise men; and to edge
A little further off from death the memory
Of any noted or bright personality
Is still a luck and poet's privilege.
And so the man who goes in my dark mind
With sand and broad waters and general kind
Of fish-and-fox-and-bird lore, and walking lank;
Knowledge of net and rod and rib and shank,
Might well stretch out my mind to be a frame—
A picture of a worthy without name.
You might see him at morning by the lock-gates,
Or busy in the warehouse on a multitude
Of boat fittings, net fittings; copper, iron, wood,
Then later digging, furious, electric
Under the apple boughs, with a short stick,
Burnt black long ages, of pipe between set teeth,
His eyes gone flaming on the work beneath—
He up-and-down working like a marionette.
Back set, eyes set, wrists; and the work self-set.

His afternoon was action but all nebulous
Trailed over four miles country, tentaculous
Of coalmen, farmers, fishermen his friends,
And duties without beginnings and without ends.
There was talk with equals, there were birds and fish
     to observe,
Stuff for a hundred thoughts on the canal's curves,
A world of sight—and back in time for tea;
Or the tide's change, his care, or a barge to let free.
The lowering of the waters, the quick inflow,
The trouble and the turmoil; characteristic row
Of exits or of river entrances;
With old (how old?) cries of the straining crews,
(Norse, Phoenician, Norse, British? immemorial use).
Tins would float shining at three-quarter tide
Midstream his line of fire, never far wide—
Dimples of water showed his aim a guide,
And ringed the sunset colours with bright ripples.

Later, tide being past violence, the gates known safe,
He would leave his station, lock warehouse and half
Conscious of tiredness now, moving lankly and slow,
Would go in a dark time like some phantom or wraith,
Most like a woodsman in full summer glow.
There he was not known to me, but as hearers know
Outside the blue door facing the canal path;
Two hours or three hours of talk; as the fishers know
Or sailors, or poachers, or wandering men know talk.

Poverty or closing time would bring him again.
On the cinder path outside would be heard his slow walk.
It had a width, that Severn chimney-corner,
A dignity and largeness which should make grave
Each word or cadence uttered or let fall, save
When the damp wind in garden shrubs was mourner.
It would have needed one far less sick than I
To have questioned, to have pried each vein of his wide
        lore.
One should be stable, and be able for wide views,
Have knowledge, and skilled manage of questions use
When the captain is met, the capable in use,
The pictured mind, the skilled one, the hawk-eyed one;
The deft-handed, quick-moving, the touch-commanded one.
Man and element and animal comprehending
And all-paralleling one. His knowledge transcending
Books, from long vain searches of dull fact.
Conviction needing instant change to act.

The nights of winter netting birds in hedges;
The stalking wild-duck by down-river sedges:
The tricks of sailing; fashions of salmon-netting:
Cunning of practice, the finding, doing, the getting—
Wisdom of every various season or light—
Fish running, tide running, plant learning and bird flight.
Short cuts, and watercress beds, and all snaring touches,
Angling and line laying and wild beast brushes;
Badgers, stoats, foxes, the few snakes, care of ferrets,
Exactly known and judged of on their merits.
Bee-swarming, wasp-exterminating and bird-stuffing.

There was nothing he did not know; there was nothing,
     nothing.

Some men are best seen in the full day shine,
Some in half-light or the dark star-light fine:
But he, close in the deep chimney-corner, seen
Shadow and bright flare, saturnine and lean;
Clouded with smoke, wrapped round with cloak of
     thought,
He gave more of desert to me—more than I ought—
Who was more used to book-poring than bright life.

One had seen half-height covering the stretched sand
With purpose, insistent, creeping-up with silver band,
But dark determined, making wide on and sure.

So behind talk flowed the true spirit—to endure,
To perceive, to manage, to be skilled to excel, to
     comprehend;

A net of craft of eye, heart, kenning and hand.
Thousand-threaded tentaculous intellect
Not easy on a new thing to be wrecked—
Since cautious with ableness, and circumspect
In courage, his mind moved to a new stand,
And only with full wisdom used that hand.

Months of firelight and lamplight of night-times; before-bed
Revelations; a time of learning and little said
On my part, since the Master he was so wise—
Easy the lesson; while the grave night-winds' sighs
At window or up chimney incessant moaning
For dead daylight or for music or fishermen dead.
Dark river voice below heard and lock's overflow.

Longford Dawns

Of course not all the watchers of the dawn
See Severn mists like forced-march mists withdrawn;
London has darkness changing into light
With just one quarter-hour of any weight.

Casual and common is the wonder grown—
Time's duty to lift light's curtain up and down,
But here Time is caught up clear in Eternity,
And draws as breathless life as you or me.

London

Clear lamps and dim stars,
Worry of heart-ache,
Are poor things should make
A consolation for the trees and bars
Of cloud on stars,
That men in western meadows stay awake
And fight with sleep for, till the strong dawn wake.

Time to Come

They will walk there, the sons of our great-grandsons and
Will know no reason for the old love of the land.
There will be no tiny bent-browed houses in the
Twilight to watch, nor small shops of multi-miscellany.
The respectable and red-brick will rule all,
With green-paint railings outside the front door wall,
And children will not play skip-games in the gutter,
Nor dust fly furious in hot valour of footer;
Queerness and untidiness will be smoothed out
As any steam-roller tactful, and there'll be no doubt
About the dustbins or the colour of curtains,
No talking at the doors, no ten o'clock flirtings,
And Nicholas will look as strange as any
Goddess ungarmented in that staid company,
With lovely attitude of fixèd grace,
But naked and embarrassed in the red brick place.

We see her well, and should have great thanksgiving,
Living in sight and form of more than common living.
She is a City still and the centuries drape her yet;
Something in the air or light cannot or will not forget
The past ages of her, and the toil which made her,
The courage of her, the army that made not afraid her,

And a shapely fullness of being drawn maybe from the air
Crystal or mellow about her or above her ever:
Record of desire apparent of dreamer or striver,
And still the house between the Cotswolds bare
And the Welsh wars; mistress of the widening river.

Midnight

There is no sound within the cottage now,
But my pen and the sound of long rain
Heavy and musical, I must think again
To find so sweet a noise, and cannot anyhow.

The soothingness and deep-toned tinkle, soft
Happenings of night, in pain there's nothing better
Save tobacco, or long most-looked-for letter . . .
The different roof-sounds—house, shed, loft and scullery.

Song

O were there anything by half
As fair as promise true
We should not change for any strange
Or violent fancy new.

But since the cheats are only real
And truths like vapours fade,
Our best advance is toward black fancies
Blind groping in dark shade.

Behind the Line

I suppose France this morning is as white as here
High white clouds veiling the sun, and the mere
Cabbage fields and potato plants lovely to see,
Back behind at Robecq there with the day free.

In the estaminets I suppose the air as cool, and the floor
Grateful dark red; the beer and the different store
Of citron, grenadine, red wine as surely delectable
As in Nineteen Sixteen; with the round stains on the dark table.

Journals Français tell the same news and the queer
Black printed columns give news, but no longer the fear
Of shrapnel or any evil metal torments.
High white morning as here one is sure is on France.

We Who Praise Poets

We who praise poets with our labouring pen
And justify ourselves with laud of men,
Have not the right to call our own our own,
Being but the groundsprouts from those great trees
    grown.
The crafted art, the smooth curve, and surety
Come not of nature till the apprentice free
Of trouble with his tools, and cobwebbed cuts,
Spies out a path his own and casts his plots.
Then looking back on four-square edifices
And wind-and-weather-standing tall houses
He stakes a court, and tries his unpaid hand,
Begins a life whose salt is arid sand,
Whose bread of cactus comes, whose wine is clear
Being bitter water from fount all too near.
Happy if after toil he grow to worth
And praise of complete men of earlier birth,
Of happier pen and more steel-propertied
Nerves: of the able and the mighty dead.

Yesterday Lost

What things I have missed today, I know very well,
But the seeing of them each new time is miracle.
Nothing between Bredon and Dursley has
Any day yesterday's precise unpraisèd grace.
The changed light, or curve changed mistily,
Coppice, now bold cut, yesterday's mystery.
A sense of mornings, once seen, for ever gone,
Its own for ever: alive, dead, and my possession.

On the Night

On the night there are shown dim few stars timorous
And light is smothered in a cloak of fear.
Are these hills out? Then night has brooded there
Of dark things till they were no more for us.

Gone are the strict falls, there is no skyline boundary,
The stars are not resting or coming to rest.
What will dawn show? A land breathing calm of breast,
Or a frightened rook-wheeling plain once bed of the sea?

Robecq Again

Robecq had straw and a comfortable tavern
Where men might their sinews feel slowly recovering
From the march-strain, and there was autumn's translucence
In the calm air and a tang of the earth and its essence.
A girl served wine there with natural dignity
Moving as any princess from care free,
And the North French air bathed crystal the flat land
With cabbages and tobacco plants and varied culture spanned,
Beautiful with moist clarity of autumn's breath.
Lovely with the year's turning to leafless death
Robecq, the dark town at night with estaminets lit,
The outside roads with poplars, plane trees on it,
Huge dark barn with candles throwing warning flares,
Glooms steady and shifting pierced with cold flowing airs,
With dumb peace at last and a wrapping from cares.

The Hoe Scrapes Earth

The hoe scrapes earth as fine in grain as sand,
I like the swirl of it and the swing in the hand
Of the lithe hoe so clever at craft and grace,
And the friendliness, the clear freedom of the place.

And the green hairs of the wheat on sandy brown.
The draw of eyes toward the coloured town,
The lark ascending slow to a roof of cloud
That cries for the voice of poetry to cry aloud.

Old Dreams

Once I had dreamed of return to a sunlit land,
Of summer and firelight winter with inns to visit,
But here are tangles of fate one does not understand,
And as for rest or true ease, where is it or what is it?

With criss-cross purposes and spoilt threads of life,
Perverse pathways, the savour of life is gone.
What have I then with crumbling wood or glowing coals,
Or a four-hours' walking, to work, through a setting sun?

The Escape

I believe in the increasing of life: whatever
Leads to the seeing of small trifles,
Real, beautiful, is good; and an act never
Is worthier than in freeing spirit that stifles
Under ingratitude's weight, nor is anything done
Wiselier than the moving or breaking to sight
Of a thing hidden under by custom—revealed,
Fulfilled, used (sound-fashioned) any way out to delight:
Trefoil—hedge sparrow—the stars on the edge at night.

On Foscombe Hill

O exquisite
And talking water, are you not more glad
To be sole daughter and one comfort bright
Of this small hill lone-guarding its delight,
Than unconsidered to be
Some waif of Cotswold or the Malvern height?
Your name a speck of glory in so many.
You are the silver of a dreaming mound
That likes the quiet way of thought and sound,
Moists tussocks with a sunken influence,
Collects and runs one way down to farm yard
Sheds, house, standing up there by soft sward,
Green of thorn, green of sorrel and age-old heath
Of South-West's lovely breath.

Friendly Are Meadows

Friendly are meadows when the sun's gone in
And no bright colour spoils the broad green of grey,
And one's eyes rest looking to far Cotswold away
Under cloud ceilings whorled and most largely fashioned
With seventeenth-century curves of the tombstone way.
A day of softnesses, of comfort of no din, not passioned.
Sorrel makes rusty rest for the eyes, and the worn path,
Brave elms, and stiles, willows by dyked deep water-run—
North French general look, and a sort of bath
Of freshness—a light wrap of comfortableness
Over one's being, a sense as of music begun;
A slow gradual symphony of worthiness, fulfilledness.
But this is Cotswold, Severn: when these go stale
Then the all-universal and wide decree shall fail
Of world's binding, and earth's dust apart be loosed,
And man's worship of all grey comforts be abused,
To mere wonder at lightning and torrentous strong flying hail.

The Bare Line of the Hill

The bare line of the hill
Shows Roman and
A sense of Rome hangs still
Over the land.

So that one looks to see
Steel gleam, to hear
Voices outflung suddenly
Of the challenger.

Yet boom of the may-fly
The loudest thing
Is of all under the sky
Of the wide evening.

And the thing metal most
The pond's last sheen
Willow shadow crossed
But still keen.

How long, how long before
The ploughland lose
Sense of that old power?
The winds, the dews

Of twice ten hundred years
Have dimmed no jot
Of Roman thought there, fears,
Triumphs unforgot.

Has Caesar any thought
In his new place, of lands
Far west, where cohorts fought,
Watched at his commands?

Carausius, Maximus,
Is all let slip, then why
Does Rome inherit thus
Dominate memory

So royally that Here
And Now are nothing known?
The regal and austere
Mantle of Rome is thrown

As of old—about the walls
Of hills and the farm—the fields.
Scabious guards the steeps,
Trefoil the slopes yield.

Had I a Song

Had I a song
I would sing it here
Four lined square shaped
Utterance dear

But since I have none,
Well, regret in verse
Before the power's gone
Might be worse, might be worse.

Quiet Fireshine

Quiet is fireshine when the light is gone,
The kettle's steam is comfort and the low song.
Now all the day's business stills down and is done
To watch them seems but right; nothing at all is wrong.

Save the dark thoughts within most bitter with
Disappointment, mere pain; gnawing at heart's peace.
The heavy heart so ponderous once was lithe
Travelling the hill slope easy at light's pace.

Quiet is fireshine, and the mind would soak
Years ago, after football, in drowse of light.
Now the slack body is sick and a bitter joke
To a soul too sick for dreams at fall of night.

The High Hills

The high hills have a bitterness
Now they are not known
And memory is poor enough consolation
For the soul hopeless gone.
Up in the air there beech tangles wildly in the wind—
That I can imagine
But the speed, the swiftness, walking into clarity,
Like last year's bryony are gone.

Old Thought

Autumn, that name of creeper falling and tea-time loving,
Was once for me the thought of high Cotswold noon-air,
And the earth smell, turning brambles, and half-cirrus
     moving,
Mixed with the love of body and travel of good turf there.

O up in height, O snatcht up, O swiftly going,
Common to beechwood, breathing was loving, the yet
Unknown Crickley Cliffs trumpeted, set music on glowing
In my mind. White Cotswold, wine scarlet woods and leaf
     wreckage wet.

Kettle Song

The worry and low murmur
Of the black kettle are set
Against my unquiet achings
And vanish, so strong is the fret.

Such tangles and evil-skeined fibres
Of living so matted are grown
That water-song is hardly noticed
For all its past comfortings known.

Hedges

'Bread and cheese' grow wild in the green time,
Children laugh and pick it, and I make my rhyme
For mere pleasure of seeing that so subtle play
Of arms and various legs going every, any way.

And they turn and laugh for the unexpensiveness
Of country grocery and are pleased no less
Than hedge sparrows. Lessons will be easier taken,
For this gipsy chaffering, the hedge plucked and shaken.

When March Blows

When March blows, and Monday's linen is shown
On the gooseberry bushes, and the worried washer alone
Fights at the soaked stuff, meres and the rutted pools
Mirror the wool-pack clouds and shine clearer than jewels.

And the children throw stones in them, spoil mirrors and clouds.
The worry of washing over, the worry of foods
Brings tea-time; March quietens as the trouble dies.
The washing is brought in under wind-swept clear infinite skies.

Song

Past my window dawn and down
   Through the open shutters thrown
Pass the birds the first awaking,
   And the light wind peace breaking.

Now the ink will dry on pen—
   And the paper take no more
Thoughts of beauty from the far
   Night, or remembered day of men.
Cotswold breaking the dark or standing
   Brave as the sun, with white scar.

Now my footsteps shall go light
   By the fence and bridge till white
The farm show, that till now had glimmered.
   In the trees July had summered.

George Chapman—The Iliad

The football rush of him, and that country knowledge,
That pluck driving through work, endless that wearying
    courage,
Still unwearying. Still face on: and the wide heaven taking
At one glance in at his eye: O that set on shaking
Keats, and new wonder brought fresh from mortal power.
That first hand still, and in sad heart-break hour
The voice of David brought once again to say
What joy what grief on man Time's heavy hand doth lay.
Homer's re-bringer, and of joy that great map-man,
Friend of great makers: scholar, mixed-minded doer, George
    Chapman.

The Cloud

One could not see or think, the heat overcame one,
With a dazzle of square road to challenge and blind one,
No water was there, cow-parsley the only flower
Of all May's garland this torrid before-summer hour,
And but one ploughman to break ten miles of solitariness.
No water, water to drink, stare at, the lovely clean grained one.

Where like a falcon on prey, shadow flung downward
Solid as gun-metal, the eyes sprang sunward
To salute the silver radiance of an Atlantic high
Prince of vapour required of the retinue
Continual changing of the outer-sea's flooding sun.
Cloud royal, born called and ordered to domination,
Spring called him out of his tent in the azure of pleasure,
He girt his nobleness—and in slow pace went onward
A true monarch of air chosen to service and station,
And directed on duties of patrolling the considered blue.
But what his course required being fulfilled, what fancy
Of beyond-imagination did his power escape to
With raiment of blown silver . . .

Early Spring Dawn

Long shines the thin light of the day to north-east,
The line of blue faint known and the leaping to white;
The meadows lighten, mists lessen, but light is increased,
The sun soon will appear, and dance, leaping with light.

Now milkers hear faint through dreams first cockerel crow,
Faint yet arousing thought, soon must the milk pails be flowing.
Gone out the level sheets of mists, and see, the west row
Of elms are black on the meadow edge. Day's wind is blowing.

Clay

The clay our mothers feed us with is taken
To be the tie and case of the bright spirit,
It is washed and dressed and hindered and does inherit
A thousand vermin cares are of all Time's making.
The clay does enter, does possess us and we have then
A thousand clayey consequences, known
By the hurt and hinders of action, and the left undone
Adventurous things which salt the lives of men.
That clay has cased the fibre and tied the limb
Nor yet that entail shall we put away
Till the clay wrap us affectionately in clay
And the adhesive marl film the bright eyeglance dim.

Larches

Larches are most fitting small red hills
That rise like swollen ant-heaps likeably
And modest before big things like near Malvern
Or Cotswold's further early Italian
Blue arrangement; unassuming as the
Cowslips, celandines, buglewort and daisies
That trinket out the green swerves like a child's game.
O never so careless or lavish as here.
I thought, 'You beauty! I must rise soon one dawn time
And ride to see the first beam strike on you
Of gold or ruddy recognisance over
Crickley level or Bredon sloping down.
I must play tunes like Burns, or sing like David,
A saying-out of what the hill leaves unexprest,
The tale or song that lives in it, and is sole,
A round red thing, green upright things of flame'.
It is May, and the conceited cuckoo toots and whoos
    his name.

The Bronze Sounding

In the old days autumn would clang a gong
Of colour between Cranham and the Birdlip curve,
Hollow brass sustained the woods' noble swerve
And the air itself stood against music as crystal strong.
So it may be so still, but the body now
No longer takes in distance as slow thought.
Old man's beard may be tangled in black hedges caught,
But the body hurt, spirit is hindered and slow,
And evil hurts me past my maker's right.

February Dawn

Rooks flew across the sky, bright February watched
Their steady course straight on, like an etcher's line scratched.
The dark brown or tawny earth breathed incense up,
I guessed there were hidden daisies, hoped the first buttercup.

The tunes of all the county, old-fashioned and my own
Wilful, wanton, careless, thronged in my mind, alone.
The sight of earth and rooks made passion rise in my blood.
Far gleamed Cotswold. Near ran Severn. A god's mood.

Save that I knew no high things would amaze day-fall
I had prayed heaven to kill me at that time most to fulfil
My dreams for ever. But looked on to a west bright at five,
Scarred by rooks in purpose; and the late trees in strife.

Moments

I think the loathed minutes one by one
That tear and then go past are little worth
Save nearer to the blindness to the sun
They bring me, and the farewell to all earth

Save to that six-foot-length I must lie in
Sodden with mud, and not to grieve again
Because high autumn goes beyond my pen
And snow lies inexprest in the deep lane.

The Dark Tree

Strange that the dark tree, the unblossomed apple
Should show so much the princess of this orchard,
Whose clear shadows equal the pear's dappled,
And patterns black for cloudy the smooth sward.
The light-bearing tree should far surpass
This slim and winter-garmented young slut,
That's never watched her dark fire in the glass,
Nor wondered vainly till her eyelids shut.
The Cinderella of unjealous kin
Who watch their sister till her time comes in.

Felling a Tree

The surge of spirit that goes with using an axe,
The first heat—and calming down till the stiff back's
Unease passed, and the hot moisture came on body.
There under banks of Dane and Roman with the golden
Imperial coloured flower, whose name is lost to me—
Hewing the trunk desperately with upward strokes;
Seeing the chips fly—(it was at shoulder height, the trunk)
The green go, and the white appear—
Who should have been making music, but this had to be done
To earn a cottage shelter, and milk, and a little bread:
To right a body, beautiful as water and honour could make
    one—
And like the soldier lithe of body in the foremost rank.
I stood there, muscle stiff, free of arm, working out fear.
Glad it was the ash tree's hardness not of the oaks', of the iron
    oak.
Sweat dripped from me—but there was no stay and the echoing
    bank
Sent back sharp sounds of hacking and of true straight
    woodcraft.
Some Roman from the pinewood caught memory and laughed.
Hit, crack and false aim, echoed from the amphitheatre
Of what was Roman before Romulus drew shoulder of Remus
Nearer his own—or Fabius won his salvation of victories.
In resting I thought of the hidden farm and Rome's hidden mild
    yoke
Still on the Gloucester heart strong after love's fill of centuries,
For all the happy, or the quiet, Severn or Leadon streams.
Pondered on music's deep truth, poetry's form or metre.
Rested—and took a thought and struck onward again,
Who had frozen by Chaulnes out of all caring of pain—
Leant Roman fortitude at Laventie or Ypres.
Saw bright edge bury dull in the beautiful wood,
Touched splinters so wonderful—half through and soon to come
    down
From that ledge of rock under harebell, the yellow flower—the
    pinewood's crown.
Four inches more—and I should hear the crash and great thunder
Of an ash Crickley had loved for a century, and kept her own.

Thoughts of soldier and musician gathered to me—
The desire of conquest ran in my blood, went through me—
There was a battle in my spirit and my blood shared it—
Maisemore—and Gloucester—bred me, and Cotswold reared it,
This great tree standing nobly in the July's day full light
Nearly to fall—my courage broke—and gathered—my breath
     feared it,
My heart—and again I struck, again the splinters and steel glinters
Dazzled my eyes—and the pain and the desperation and near
     victory
Carried me onwards—there were exultations and mockings
     sunward.
Sheer courage, as of boat sailings in equinoctial unsafe squalls,
Stiffened my virtue—and the thing was done. No. Dropped my
     body—
The axe dropped—for a minute, taking breath, and gathering
     the greedy
Courage—looking for rest to the farm and grey loose-piled
     walls—
Rising like Troilus to the first word of 'Ready'—
The last desperate onslaught—took the two inches of too steady
Trunk—on the rock edge it lurched, threatening my labouring life
(Nearly on me). Like Trafalgar's own sails imperiously moving
     to defeat
Across the wide sky unexpected glided and the high bank's pines
     and fall straight,
Lower and lower till the crashing of the fellow trees made strife.
The thud of earth, and the full tree lying low in state,
With all its glory of life and sap quick in the veins . . .
Such beauty, for the farm fires and heat against chilly rains,
Golden glows in the kitchen from what a century made great . . .

The axe fell from my hand, and I was proud of my hand.
Crickley forgave, for her nobleness, the common fate of trees.
As noble or more noble, the oak, the elm that is treacherous,
But dear for her cherishing to this beloved and this rocky land.
Over above all the world there, in a tired glory swerved there—
To a fall, the tree that for long had watched Wales glow strong,
Seen Severn, and farm, and Brecon, Black Mountains times
     without reckon.
And tomorrow would be fuel for the bright kitchen—for brown
     tea, against cold night.

Strange Hells

There are strange hells within the minds war made
Not so often, not so humiliatingly afraid
As one would have expected—the racket and fear guns made.
One hell the Gloucester soldiers they quite put out:
Their first bombardment, when in combined black shout
Of fury, guns aligned, they ducked lower their heads
And sang with diaphragms fixed beyond all dreads,
That tin and stretched-wire tinkle, that blither of tune:
'Après la guerre fini', till hell all had come down,
Twelve-inch, six-inch, and eighteen pounders hammering
    hell's thunders.

Where are they now, on state-doles, or showing shop-patterns
Or walking town to town sore in borrowed tatterns
Or begged. Some civic routine one never learns.
The heart burns—but has to keep out of face how heart burns.

The Songs I Had

The songs I had are withered
Or vanished clean,
Yet there are bright tracks
Where I have been,

And there grow flowers
For others' delight.
Think well, O singer,
Soon comes night.

Smudgy Dawn

Smudgy dawn scarfed with military colours
Northward, and flowing wider like slow sea water,
Woke in lilac and elm and almost among garden flowers.
Birds a multitude; increasing as it made lighter.
Nothing but I moved by railings there; slept sweeter
Than kings the country folk in thatch or slate shade.
Peace had the grey west, fleece clouds sure in its power—
Out on much-Severn I thought waves readied for laughter,
And the fire-swinger promised behind the elm-pillars
A day worthy such beginning to come after.
To the room then to work with such hopes as may
Come to the faithful night worker, in west country's July.

Lovely Playthings

Dawn brings lovely playthings to the mind,
But sunset fights and goes down in battle blind.
The banners of dawn spread over in mystery,
But nightfall ends a boast and a pageantry.

After the halt of dawn comes the slow moving of
Time, till the sun's hidden rush and the day is admitted.
Sunset dies out in a smother of something like love,
With dew and the elm-hung stars and owl outcries
    half-witted.

The Not-Returning

Never comes now the through-and-through clear
Tiredness of body on crisp straw down laid,
Nor the tired thing said
Content before the clean sleep close the eyes,
Or ever resistless rise
Pictures of far country westward, westward out of the sight
    of the eyes.
Never more delight comes of the roof dark lit
With under-candle flicker nor rich gloom on it,
The limned faces and moving hands shuffling the cards,
The clear conscience, the free mind moving towards
Poetry, friends, the old earthly rewards.
No more they come. No more.
Only the restless searching, the bitter labour,
The going out to watch stars, stumbling blind through the
    difficult door.

What Evil Coil

What evil coil of fate has fastened me
Who cannot move to sight, whose bread is sight,
And in nothing has more bare delight
Than dawn or the violet or the winter tree.
Stuck-in-the-mud—blinkered-up, roped for the fair.
What use to vessel breath that lengthens pain?
O but the empty joys of wasted air
That blow on Crickley and whimper wanting me!

Swift and Slow

Death swooped suddenly on men in Flanders
There were no tweedledees or handy-danders
The skull was cleft, the life went out from it
And glory in a family tale was set.
But here, having escaped the steely showers
Endured through panged intolerable hours
The expensive and much determined doom,
Find slow death in the loved street and bookish room.
Liver and bowels congested to devil's pitch
For a pittance or sake of benefit, what matters which?
Life witch-like seen as Dürer saw, the detested witch.

When the Body Might Free

When the body might free, and there was use in walking,
In October time—crystal air-time and free words were talking
In my mind with light tunes and bright streams ran free;
When the earth smelt, leaves shone and air and cloud had glee;

Then there was salt in life but now none is known
To me who cannot go either where the white is blown
Of the grass, or scarlet willow-herb of past memory.
Nothing is sweet to thinking, nothing from life free.

In the Old Time

In the old time when September's stubble gleamed
And as the content of all folk-writing seemed
The true consolation for all woes, I made
Music out of stubbornness and was glad.
But now the pen writes words, and the brain is content,
Fates haggle for me, the body has its bent,
And only theological and ethical discussions
Continue like a toothache, from black hidden dread.

Sonnet—September 1922

Fierce indignation is best understood by those
Who have time or no fear, or a hope in its real good.
One loses it with a filed soul or in sentimental mood.
Anger is gone with sunset, or flows as flows
The water in easy mill-runs; the earth that ploughs
Forgets protestation in its turning, the rood
Prepares, considers, fulfils; and the poppy's blood
Makes old the old changing of the headland's brows.

But the toad under the harrow toadiness
Is known to forget, and even the butterfly
Has doubts of wisdom when that clanking thing goes by
And's not distressed. A twisted thing keeps still—
That thing easier twisted than a grocer's bill—
And no history of November keeps the guy.

There Is a Man

There is a man who has swept or rubbed a floor
This morning crying in the Most Holy Name
Of God for pity, and has not been able to claim
A moment's respite, that for one hour, or more.
But can the not-conceiving heart outside
Believe the atmosphere that hangs so heavy
And clouds the torment. Afterwards in the leavy
And fresher air other torments may abide,
Or pass; and new pain; but this memory
Will not pass, it is too bad and the grinding
Remains, and what is better in the finding
Of any ease from working or changing free
Words between words, and cadences in change.
But the pain is in thought, which will not freely range.

The Incense Bearers

Toward the sun the drenched May-hedges lift
White rounded masses like still ocean-drift,
And day fills with heavy scent of that gift.

There is no escaping that full current of thick
Incense; one walks, suddenly one comes quick
Into a flood of odour there, aromatic,

Not English; for cleaner, sweeter, is the hot scent that
Is given from hedges, solitary flowers, not
In mass, but lonely odours that scarcely float.

But the incense bearers, soakers of sun's full
Powerfulness, give out floods unchecked, wonderful
Utterance almost, which makes no poet grateful,

Since his love is for single things rarely found,
Or hardly: violets blooming in remote ground,
One colour, one fragrance, like one uncompanied sound

Struck upon silence, nothing looked-for. Hung
As from gold wires this May incense is swung,
Heavy of odour, the drenched meadows among.

There Have Been Anguishes

There have been anguishes
In the different poetries
Where the man's mind cries
Out on God's deep mercies.
None has denied them,
They are of old time
And a faded rhyme
No living one does condemn.

But half my suffering,
Told out in pencil or
Ink as night came, before
Justice or wītan-ring,
Would not gain redress
For its strange seeming.
And a true deeming
Lacked of its witness.

Vain is the use of the mind,
Almost the soul halts here,
Consumed with black fear,
Black fear of a pain-blind
Nature, that craves ending
To such bad being,
Or truly to be seeing
At least the use and mending.

Riez Bailleul

Behind the line there mending reserve posts, looking
On the cabbage fields with other men carefully tending
    cooking;
Hearing the boiling; and being sick of body and heart,
Too sick for anything but hoping that all might depart—
We back in England again, and white roads to walk on,
Eastwards to hill-steeps, or see meadows good to go talk on.
Grey Flanders sky over all and a heaviness felt
On the sense that no working or dreaming would any way
    melt . . .
This is not happy thought, but a glimpse most strangely
Forced from the past, to hide this pain and work myself free
From present things. The parapet, the grey look-out, the
    making
Of a peasantry, by dread war, harried and set on shaking;
A hundred things of age, and of carefulness,
Spoiling; a farmer's treasure perhaps soon a wilderness.

Old Times

Out in the morning
For a speed of thought I went,
And a clear thought of scorning
For homekeeping; while downward bent
Grass blades with dewdrops
Heavy on those delicate
Sword shapes, wonder thereat
Brightening my first hopes.

A four hours' tramping
With brisk blood flowing,
And life worth knowing
For all that something
Which let happiness then—
Sometimes, not always,
Breath-on-mirror of days—
And all now gone, since when?

To God

Why have you made life so intolerable
And set me between four walls, where I am able
Not to escape meals without prayer, for that is possible
Only by annoying an attendant. And tonight a sensual
Hell has been put on me, so that all has deserted me
And I am merely crying and trembling in heart
For death, and cannot get it. And gone out is part
Of sanity. And there is dreadful hell within me.
And nothing helps. Forced meals there have been and electricity
And weakening of sanity by influence
That's dreadful to endure. And there is Orders
And I am praying for death, death, death,
And dreadful is the indrawing or out-breathing of breath
Because of the intolerable insults put on my whole soul,
Of the soul loathed, loathed, loathed of the soul.
Gone out every bright thing from my mind.
All lost that ever God himself designed.
Not half can be written of cruelty of man, on man,
Not often such evil guessed as between man and man.

On Somme

Suddenly into the still air burst thudding
And thudding, and cold fear possessed me all,
On the grey slopes there, where winter in sullen brooding
Hung between height and depth of the ugly fall
Of Heaven to earth; and the thudding was illness' own.
But still a hope I kept that were we there going over,
I in the line, I should not fail, but take recover
From others' courage, and not as coward be known.
No flame we saw, the noise and the dread alone
Was battle to us; men were enduring there such
And such things, in wire tangled, to shatters blown.
Courage kept, but ready to vanish at first touch.
Fear, but just held. Poets were luckier once
In the hot fray swallowed and some magnificence.

A Wish

I would hope for the children of West Ham
Wooden-frame houses square, with some-sort stuff
Crammed in to keep the wind away that's rough,
And rain; in summer cool, in cold comfortable enough.
Easily destroyed—and pretty enough, and yet tough.
Instead of brick and mortar tiled houses of no
Special appearance or attractive show.
Not crowded together, but with a plot of land,
Where one might play and dig, and use spade or the hand
In managing or shaping earth in such forms
As please the sunny mind or keep out of harms
The mind that's always good when let go its way
(I think) so there's work enough in a happy day.

Not brick and tile, but wood, thatch, walls of mixed
Material, and buildings in plain strength fixed.
Likeable, good to live in, easily pulled
Down, and in winter with warm ruddy light filled—
In summer with cool air; O better this sort of shelter—
And villages on the land set helter-skelter
On hillsides, dotted on plains—than the too exact
Straight streets of modern times, that strait and strict
And formal keep man's spirit within bounds,
Where too dull duties keep in monotonous rounds.

These villages to make for these towns of today—
O haste—and England shall be happy with the May
Or meadow-reach to watch, miles to see and away.

Hazlitt

Hazlitt, also, tea-drinker and joyous walker,
To him we give thanks and are grateful that
He saw Shakespeare as man, not as over-great
Above the comprehending of reader and talker.
In swift English writing of Hamlet and those
Live-moving figures of poetry and high wit
He cast new wholesome light on the sense that writ
And a love that with friendship is instinct and glows.
Never more need of brave minds than in setting free
This figure of Time for friendship and not for fear:
Shakespeare, companier of men that were lively here
And walked happy in thought in an England Merry.
So to Hazlitt thanks, and a clear thought of gratitude,
For his speaking in words so plain our Shakespeare's mood.

Hedger

To me the A Major Concerto has been dearer
Than ever before, because I saw one weave
Wonderful patterns of bright green, never clearer
Of April; whose hand nothing at all did deceive
Of laying right
The stakes of bright
Green lopped-off spear-shaped, and stuck notched, crooked-up;
Wonder was quickened at workman's craftsmanship
But clumsy were the efforts of my stiff body
To help him in the laying of bramble, ready
Of mind, but clumsy of muscle in helping; rip
Of clothes unheeded, torn hands. And his quick moving
Was never broken by any danger, his loving
Use of the bill or scythe was most deft, and clear—
Had my piano-playing or counterpoint
Been so without fear
Then indeed fame had been mine of most bright outshining;
But never had I known singer or piano-player
So quick and sure in movement as this hedge-layer
This gap-mender, of quiet courage unhastening.

Memory

They have left me little indeed, how shall I best keep
Memory from sliding content down to drugged sleep?
But my blood, in its colour even, is known fighter.
If I were hero for such things here would I make wars
As love for dead things trodden under in January's stars,

Or the gold trefoil itself spending in careless places
Tiny graces like music's for its past exquisitenesses.
Why war for huge domains of the planet's heights or plains?
(Little they leave me.) It is a dream. Hardly my heart dares
Tremble for glad leaf-drifts thundering under January's stars.

Song

I had a girl's fancies
At the pools—
And azure at chances
In the rut holes
Minded me of Maisemore,
And Gloucester men
Beside me made sure
All faithfulness again.

The man's desiring
Of great making
Was denied, and breaking
To the heart, much-caring—
Only the light thoughts
Of the poet's range
Stayed in that war's plights.
Only, soul did not change.

So, to the admiration
Of the rough high virtues
Of common marching
Soldiers, and textures
Of russet noblenesses,
My mind was turned.
But where are such verses
That in my heart burned?

The Mangel-Bury

It was after war; Edward Thomas had fallen at Arras—
I was walking by Gloucester musing on such things
As fill his verse with goodness; it was February; the long
    house
Straw-thatched of the mangels stretched two wide wings;
And looked as part of the earth heaped up by dead soldiers
In the most fitting place—along the hedge's yet-bare lines.
West spring breathed there early, that none foreign divines.
Across the flat country the rattling of the cart sounded;
Heavy of wood, jingling of iron; as he neared me I waited
For the chance perhaps of heaving at those great rounded
Ruddy or orange things—and right to be rolled and hefted
By a body like mine, soldier still, and clean from water.
Silent he assented; till the cart was drifted
High with those creatures, so right in size and matter.
We threw with our bodies swinging, blood in my ears
    singing;
His was the thick-set sort of farmer, but well-built—
Perhaps, long before, his blood's name ruled all,
Watched all things for his own. If my luck had so willed
Many questions of lordship I had heard him tell—old
Names, rumours. But my pain to more moving called
And him to some barn business far in the fifteen acre field.

The Dream

There had been boat-sailing on Severn river,
And when London was reached, it seemed most easy—
Of right—to look for such joy as to see sails quiver
And pull the rudder hard round, against the breezy
Wind out of Essex, or off Kentish shores.
So to Rotherhithe blue as to dancing water,
Seeing the cleaving water before prow scatter,
And the moving surface so wonderful like bright floors.
And doubtful of all things, asked an owner there
Whether a boat might be had cheap, but little hoped
Since money was not mine, and such chance escaped
Any but those with twenty pounds to spare.
It was worrying a good man, but there was that one
Hope in me of getting a sail up, to see foam run.

War Poet

I know that honour is
Because I follow it.
I know that love is
My heart does cry for it.

The sun? I dare not watch.
The stars? I was night-walker:
My friends in the high arch—
By Cranham or high Crickley

They hurt like unsought kisses
From a love one dare
Not love—they are the water-hisses
From a cooled iron, red-bare.

Greatness? I have sailed
A boat in March daring ...
And made a music, called
All March to my caring

Whether I made well
Or no—and Vermand knows
Colour of my blood—Neuve Chapelle
Courage—as war's courage goes.

Love? A hundred know it.
Men have seen my eyes.
Women have watched love, though it
Failed at actualities.

Steel-bound to my service
Earth, blood and all.
Only England refuses ...
Only life does not call ...

Only meanness hurts her heart
Only rust her steel ...
Only ... She is coward, coward ...
And I suffer agonies, rightly unheard,
Because she likes sin too well.

The Depths

Here no dreams touch me to colour
Sodden state of all-dolour:
No touch of peace, no creation
Felt, nor stir of divination.

Friend of stars, things, inky pages—
Knowing so many heritages
Of Britain old, or Roman newer;
Here all witchcrafts scar and skewer.

Coloured maps of Europe taking
And words of poets fine in making,
I march once more with hurt shoulders,
And scent the air, a friend with soldiers.

Devil's doom that none guess,
Evil's harms worshipped no less,
Grind my soul—and no god clutches
Out of darks god's-honour smutches.

While I Write

While I write war tells me truth; as for brave
None might challenge Gloucesters, save those dead who have
Paid prices for pre-eminence, perhaps have got their pay.
But the common goodness of those soldiers shown day after day,
And the sight of each-hour beauty brilliant or most grave,
Stays with me yet. While I am forbidden to write
Tale of the continual readiness for a bad bloodiness,
And steadiness against hell-fire; and strained eyes with
    humour bright.
War told me truth: I have Severn's right of maker,
As of Cotswold: war told me: I was elect, I was born fit
To praise the three hundred feet depth of every acre
Between Tewkesbury and Stroudway, Side and Wales Gate.

It Is Winter

It is winter, the soon dark annoys me—
Who cannot remember Severn her warm dark lights;
And am too tortured to remember old ploys the
Gloucesters used to please themselves in the straits
Of poverty and idleness of French villages.
Then before opening-time they would walk house-bordered
Or leafy ways—hurrying, keeping off the fierce cold.
Then when the lights showed, the estaminet's time came,
They would hammer on the door; they would shout out
        good-mannered
Rudenesses; enter, sit within, and as careful
As old ladies of knitting would drink beer or more
        honoured
Wine, trembling at the expense, which to them was fearful;
Bask in the warm, dream poetry of the gold flame.
While the poet watching their faces, and envying noble
Poetry of Long Island—strong, human, star-bannered—
Sat also, accusing time for his music, lost in service,
        refusing all blame.

It Is Near Toussaints

It is near Toussaints, the living and dead will say:
'Have they ended it? What has happened to Gurney?'
And along the leaf-strewed roads of France many brown shades
Will go, recalling singing, and a comrade for whom also they
Had hoped well. His honour them had happier made.
Curse all that hates good. When I spoke of my breaking
(Not understood) in London, they imagined of the taking
Vengeance, and seeing things were different in future.
(A musician was a cheap, honourable and nice creature.)
Kept sympathetic silence; heard their packs creaking
And burst into song—Hilaire Belloc was all our master.
On the night of all the dead, they will remember me,
Pray Michael, Nicholas, Maries lost in Novembery
River-mist in the old City of our dear love, and batter
At doors about the farms crying 'Our war poet is lost.
Madame—no bon!'—and cry his two names, warningly,
   sombrely.

Regrets After Death

True on the Plain I might have seen Salisbury Close,
But how that would have repaid there is no one knows,
True at Epping I might have thanked kindness more,
But we were for France then—scarce a week to be here.
At Chelmsford, true I might have kept my first lodging
Despite of cooking 'cause she did my washing.
But since no more of France I saw than three
Weeks, and had no more honour of battle than the
One name, the still line of East Laventie,
Regrets and hopes and accusations are all vain.
Chelmsford was bad, Hell-upon-Army the Plain,
Epping had compensations, Northampton kindness,
    invitations.
They buried me in Artois, with no time to complain.

Butchers and Tombs

After so much battering of fire and steel
It had seemed well to cover them with Cotswold stone—
And shortly praising their courage and quick skill
Leave them buried, hidden till the slow, inevitable
Change came should make them service of France alone.
But the time's hurry, the commonness of the tale
Made it a thing not fitting ceremonial,
And so the disregarders of blister on heel,
Pack on shoulder, barrage and work at the wires,
One wooden cross had for ensign of honour and life gone—
Save when the Gloucesters turning sudden to tell to one
Some joke, would remember and say—'That joke is done,'
Since he who would understand was so cold he could not feel,
And clay binds hard, and sandbags get rotten and crumble.

The Bohemians

Certain people would not clean their buttons,
Nor polish buckles after latest fashions,
Preferred their hair long, putties comfortable,
Barely escaping hanging, indeed hardly able;
In Bridge and smoking without army cautions
Spending hours that sped like evil for quickness,
(While others burnished brasses, earned promotions).
These were those ones who jested in the trench,
While others argued of army ways, and wrenched
What little soul they had still further from shape,
And died off one by one, or became officers.
Without the first of dream, the ghost of notions
Of ever becoming soldiers, or smart and neat,
Surprised as ever to find the army capable
Of sounding 'Lights out' to break a game of Bridge,
As to fear candles would set a barn alight:
In Artois or Picardy they lie—free of useless fashions.

Autumn

Autumn, dear to walkers with your streaks and carpets
Of bright colours, spread like a boy's gift for the true boy,
Sacred for the love flowing over and unuttered even in making—
Have you too left me?

Never was trust so equal between man and his dear mates
Of tree or watercourse flowing by Cranham or past Hartpury.
Eternity promised: what unfaith could cause any shaking
In that love, near bereft me?

Earth spaces breathing dark incense (as the kind shower wets)
And woodlands stirring to blood-light, the heart all ready—
Could you not, with your untouched power, save me from
     this breaking
Tyranny; not Severn have safed me?

Signallers

To be signallers and to be relieved two hours
Before the common infantry—and to come down
Hurriedly to where estaminet's friendliest doors
Opened—where before the vulgar brawling common crew
Could take the seats for tired backs, or take the wine
Best suited for palates searching for delicate flavours
(Or pretty tints) to take from the mind trench ways and
    strain,
Though it be on tick, with delicately wangled sly favours.
Then having obtained grace from the lady of the inn—
How good to sit still and sip with all-appreciative lip,
(After the grease and skilly of line-cookhouse tea)
The cool darkling texture of the heavenly dew
Of wine—to smoke as one pleased in a house of courtesy—
Signallers, gentlemen, all away from the vulgar
Infantry—so dull and dirty and so underpaid,
So wont to get killed and leave the cautious signallers
To signal down the message that they were dead.

Anyway, distinctions or not—there was a quiet
Hour or so before the Company fours halted, and were
Formed two-deep, and dismissed and paid after leaden, dilatory
Hanging around, to bolt (eager) to find those apparently
Innocent signallers drinking, on tick, at last beer.

Portraits

Looking at Washington, Lee and Jefferson,
And the Scottish pride of Andrew Jackson;
Many more; one believes in the Nation that left
Men of so human a sort to manage the believed-
In States, through freedom's struggle, and the settling;
But why ever to trouble, much less to battling,
Since freedom was so young and hard a gift?
(Grant's face serving a bad cause on condition,
And success—and loathing.) It is best to turn
To those of Thomas Jefferson, or Dolly Madison
(So pretty), or Mrs Derby almost as beautiful
Or James McNeil Whistler's of an unknown girl
Out from Jane Austen or the Mozart Concertos,
Looking from the pages, unembarrassed, life-cool;
Or Miss Curtis of the farmhouses, dairies
Of Gloucestershire. . . . A happy, but too lazy life
When there were few books (out of the world's glorious
Many) to be had—whose goodness and whose foolishness
Shows on every page—and whose pain on one—
The Stuart portrait of General Washington.

An Appeal for Death

There is one who all day wishes to die,
And appeals for it—without a reason why—
Since Death is easy if men are merciful.
Water and land with chances are packed full.
Who all day wishes to die. How many ages
Have denied Death so—who reads old-written pages
And finds 'This man suffered and prayed for Death,
And went beyond this, desire of life beneath'?
Bitterly, bitterly, and though he feels his wrongs,
And once took pride in verse-making and in songs,
Yet now, yet now would wish to rest, and be
Out of pain, out of life, quietly, as quietly
As pained men ever were meant to rest.
Humanity knows earth to have as quiet a breast
As ever mother's to a longing child.
Therefore in mercy let rest, let rest this wild,
Or show hard torment, or of fear of such
Let rest, out of the fear of any pain's touch.
If men will not honour, nor find employ,
Will common mercy not forget what was wrong,
Remember what was good—a maker of song
Asks, desires, has prayed for mercy of Death
To end all, lie still, quiet green turf beneath,
Since promises forgotten are, and friendliness
Between so many men and him? The address
Of courtesy to casual wayfarers,
Small presents, courtesy of peace and wars—
To rest from pain, to trouble no one more—
Under green turf-mound, or by friendly shore
That will with rocking water lull his peace
That cannot now find hope nor strength nor ease.
To be let rest in mercy—to know an end
Of surety, Death's quiet surest of friend,
And what men would not, let calm Nature mend.

Snow

There's not a sound tonight.
I look out and am beaten
In my face by curious, white
Unexpected flakes
Of snow in a daze fleeting.

And retire shivering to
The warm room and the lamplight,
Where my music waits, and O
Ben Jonson lies . . .
To delight my man's nature with his great spirit.

O warmth! O golden light!
O books behind me waiting
Their turn for my love's thought . . .
Turning from work
To wrap myself in a past life of golden lighting.

Music must flow with his power, I
Bend over my task and am hard
At wrestling with the stuff for mastery
That is dumb music now—
My spirit and I wrestle, you may hear us breathing hard.

Was there ever any Love could draw me
Out of my true way of work and action?
Yes, one there was, but Time has dared show me
(A soldier and maker)
That Time dares all things, and defies ever question.

March

My boat moves and I with her delighting,
Feeling the water slide past, and watching white fashion
Of water, as she moves faster ever more whitening;
Till up at the white sail in that great sky heightening
Of fine cloth spread against azure and cloud commotion
My face looks, and there is joy in the eyes that asking
Fulfilment of the heart's true and golden passion
(Long dimmed) now gets hold of a truth and an action . . .
The ears take the sound of Severn water dashing.
The great spirit remembers Ulysses with his courage lighting
Before the danger of sea water, in a rocky passion
Of surges—and over Barrow comes the wind I've been waiting.

First Poem

O what will you turn out, book, to be?
Who are not my joy, but my escape from the worst
And most accurst of my woe? Shall you be poetry,
Or tell truth, or be of past things the tale rehearsed?

The Coin

It is hard to guess tales from the sight of a thing
Brought up suddenly to the light, though one may have blood
Of Rome, and as I, all instinct, quick to one's high mood.
So Constantine's coin suddenly upward turned here, ploughed,
Still left me dumb of word as to what the loser seemed
(Only in music my spirit rightly mused or dreamed)
And the Roman that lost this small penny-thing was most
A wonder to me, though Plutarch I had read, Virgil and others
(English). I could not get to comradeship of him, nor make ring
The coin on stone as once he might have—but stared and stood
Far-off watching the valley, the Welsh hills, with a sting
Of regret (that I, war poet, had lost this high good
Of knowing one of my infinite dead generations of brothers).
My arms might have laid friendly on his walking shoulders;
His spirit spoken to spirit of my deepest pondering . . .
So following the plough under the lovely and ancient wood
A coin was ploughed up, heating thought till it sudden grew
        ruddy and glowed.

Varennes

At Varennes also Gloucester men had their stay.
(Infantry again, of my soft job getting tired.)
Saw wonderful things of full day and of half-day:
Black pattern of twigs against the sunset dim fired;
Stars like quick inspirations of God in the seven o'clock sky.
Where the infantry drilled frozen—all all-foolishly
As on the Plain—but to the canteen went I,
Got there by high favour, having run, finished third,
In a mile race from Varennes to the next village end.
Canteen assistant, with a special care for B Company—
And biscuits hidden for favour in a manner forbidden.
Lying about chocolate to C Company hammering the gate.
Pitying them for their parade all the morning through
(Blue to the fingers, to all but the conscience blue)
Uselessly doing fatheaded things eternally.
But keeping (as was natural) Six Platoon ever in mind.
And one evening, drowsed by the wood fire I got lost in the
Blaze of warm embers, green wood smoking annoyingly;
Watched deep till my soul in the magic was rapt asleep:
Grew to power of music, and all poetries, so, uncared,
Became a maker among soldiers—dear comrades;
Which is the hardest of all wide earth's many trades;
And so proved my birthright, in a minute of warm aired
Staring into the woodfire's poetic heart, lost a tide deep.
(Until the anger of fire caught all, all in rose or gold was lost.)

Song of Autumn

Music is clear about such freshness and colour,
　　But how shall I get it?
There is great joy in walking to the quarry scar,
　　And glory—I have had it.

Beech woods have given me truest secrets, and the sighing
　　Firs I know, have told me
Truth of the hearts of children, the lovers of making.
　　Old camps have called me.

It is time I should go out to ways older than tales,
　　Walk hard, and return
To write an evening and a night through with so many wills
　　Aiding me—little now to learn.

The Last of the Book

There is nothing for me, Poetry, who was the child of joy,
But to work out in verse crazes of my untold pain;
In verse which shall recall the rightness of a former day.

And of Beauty, that has command of many gods, in vain
Have I written, imploring your help, who have let destroy
A servant of yours, by evil men birth better at once had slain.

And for my country, God knows my heart, land, men to me
Were dear there, I was friend also of every look of sun or rain;
It has betrayed as evil women wantonly a man their toy.

Soldiers' praise I had earned having suffered soldier's pain,
And the great honour of song in the battle's first grey show—
Honour was bound to me save, mine most dreadfully stain.

Rapt heart, once, hills I wandered alone, joy was comrade
    there, though
Little of what I needed was in my power; again—again
Hours I recall, dazed with pain like a still weight set to my woe.

Blood, birth, long remembrance, my County, all these have saven
Little of my being from dreadfullest hurt, the old gods have no
Pity—or long ago I should have got good, they would have
    battled my high right plain.

Of Cruelty

From the racked substance of the earth comes the plant and
That with heat and the night frost is tortured:
To some perfection that grows, man's thought wills his hand—
Roots rent, crown broken, grub-holed, it is drawn upward.

A hundred things since the first stir have hunted it,
The rooks any time might have swallowed ungrateful.
Caterpillars, slugs, as it grew, have counted on it,
And man the planter bent his gaze down on it fateful.

The thing will go to market, it must be picked up and loaded,
The salesman will doubt it or chuck it anyway in,
A horse must be harnessed first, or a donkey goaded
Before the purchaser may ever the first price pay for it.

Who may be now trembling with a vast impatience
And anxieties and mixed hopes for a resurrection
Out of the mouldering soil—to be new form, have perfections
Of flowers and petal and blade, to die, to be born to clean
  action.

For Mercy of Death

I suffer racking pain all day, and desire death so—
As few desire. Where is man's mercy gone?
Did ever past generations such torment know
Who lived near earth, and joyed when the sun shone,
Or when sweet rain came on
The earth; and the afterglow
Of sun and flowers in show
Of golden sweetness gladdened earth's dear son?
Where is that mercy now
That palpably took pleasure in the sweep
Of hedgerows—high to deep—
And houses in the making, man's own dear
Vesture and shelter here?
O sure it is that if those olden-time
Builders of farm and byre
Were here again, my pain should pity receive—
Death should make no more to grieve
My spirit with such pain it knows not how
To endure. O show
Such olden pity on poor souls in pain.
Let rest again—
As would our fathers, friends of wind and rain.

Hell's Prayer

My God, the wind is rising! on those edges
Of Cotswold dark glory might swing my soul—
And western Severn and north of water sedges
Mystery sounds, the wind's drums roll.
None will care to walk there. Those prefer to tell
Tales in a warm room of gossips, gettings, wages,
While I would be cursing exultant at the wind's toll
Of bell, shout of glory—swiftness of shadows.
My birth, my earning, my attained heritages,
Ninety times denied me now thrust so far in hell.

I think of the gods, all their old oaths and gages—
Gloucester has clear honour sworn without fail—
Companionship of meadows, high Cotswold ledges
Battered now tonight with huge wind-bursts and rages,
Flying moon glimpses like a shattered and flimsy sail—
In hell I, buried a score-depth, writing verse pages.

To Crickley

My soul goes there crying when
It is hurt by God far . . .
It is hurt too far, and moves again
By green and quarry scar.

Ages and ages dreaming there
Speak their heart to me—
Generations of tried men honour
My broken good with pity.

'Such good', they say, 'your blood had
At birth, and in this
Land was given you music in mood
Noble, true, clamorous.

And what has broken England to such
Evil is not guessed
Nor those old sentries rustling grass rough
Know, nor the rest.

Soldier that knew war's pains, poet
That kept our love—
The gods have not saved you, it is not
Our prayers lacking to move

Them to you—deep in hells now still burning
For sleep or the end's peace.
By tears we have not saved you; yearning
To accusation, and our hopes' loss, turning.
What gods are these?'

The Coppice

There is a coppice on Cotswold's edge the winds love;
It blasts so, and from below there one sees move
Tree-branches like water darkling—and I write thus
At the year's end, in nine hell-depths, with such memories;
I guess that rocks and heaves like west Irish seas;
Where the kite is this evening, that loves rock and hover
About the thin wood growth, I shall not know, cannot discover
Only guess dark ridge edge, and the gloomed valley's
Magnificence below them first night does cover.

The coppice of thin and great trees as nobly set
Against Wales for Cotswold, as it were the gate
Of Britain watching Britain, refusing ever
To acknowledge Rome; great shapes by older barrows;
That longs for me tonight as if my name were great
And owner of that swift fall, that wind-beaten swerve:
Were sayer of what the wood's heart could never forget.

Traffic in Sheets

Silk colour and lit-up candlelight the sheets I saw
By Severn's bridge that day . . .
O the lost history . . . O ladies and pageants of the mystery
Of February here and miles away.

I could have sung, but knew no fitting tunes
(For all my lore) of the spread
Of coloured sheets of the floods that ensure all June's
Dark fan-grasses of the pretty head.

The Two

Hearing the flute call
Of my country's meadows still through March blasts,
There have I hurried out and farther to the amethyst
Changes of the willows small.

And at home at night
Quiet through poetry the day's roaring shaking and rising
Me has driven to music, great mood to iron-twisting changing:
Withered leaves at next seeing.

That at least gave dawn
When to the upper windows of the house all else still
Climbed I, saw magnificent dawn-pageant of the daffodil
And rose-on-thorn come on.

Musers Afar Will Say

Soon will blow the changed leaves on the Malvern road
Hurtling in tiny charge; and the streams will be strowed
With arrows and shields of a hundred affectionate fields . . .
But who to dear desiring Gloucester will bring good?

Men afar off by strange rivers will say, 'Now the year
Will be storming and crying triumphant before death there
By Gloucester.' But who will bring Severn any heart's good,
Full-heralded bright river? They will fear, and they'll remember.

The Dancers

The dancers danced in a quiet meadow
It was winter, the soft light lit in clouds
Of growing morning—their feet on the firm
Hillside sounded like a baker's business
Heard from the yard of his beamy barn-grange.
One piped, and the measured irregular riddle
Of the dance ran onward in tangling threads . . .
A thing of the village centuries old in charm.
With tunes from the earth they trod, and naturalness
Sweet like the need of pleasure of change.
For a lit room with panels gleaming
They practised this set by winter's dreaming
Of pictures as lovely as are in spring's range . . .
No candles, but the keen dew-drops shining . . .
And only the far jolly barking of the dog strange.

I Would Not Rest

I would not rest till work came from my hand
And then as the thing grew, till fame came,
(But only in honour) . . . and then, O, how the grand
Divination of ages grew to faith's flame.
Great were our fathers and beautiful in all name,
Happy their days, lovely in considered grain each word,
Their days were kindness, growth, happiness, mindless.

I would not rest until my County were
Thronged with the Halls of Music; and until clear
Hospitality for love were e'er possible . . .
And any for honour might come, or prayer, to certain
Fondness and long nights' talking till all's known.

Madness my enemy, cunning extreme my friend,
Prayer my safeguard. (Ashes my reward at end.)
Secrecy fervid my honour, soldier-courage my aid.
(Promise and evil threatening my soul ever-afraid.)
Now, with the work long done, to the witchcraft I bend
And crouch—that knows nothing good, Hell uncaring
Hell undismayed.

The Pleasance Window

Now light dies from the pleasance
With rich look of colour
About the lawn, and winter
Whispers of leafless trees . . .

Of orange sunsets soft dying
And chill in the air at morning,
When friends will delight also
To speak of Artois' friends,

Warm straw after freezing ditches.
And soon will come long dark
Starry when the poet goes
Content with his masters out to stars.

Sea-Marge

Pebbles are beneath, but we stand softly
On them, as on sand, and watch the lacy edge
Of the swift sea

Which patterns and with glorious music the
Sands and round stones. It talks ever
Of new patterns.

And by the cliff-edge, there, the oakwood throws
A shadow deeper to watch what new thing
Happens at the marge.

What Was Dear

What was dear to Pan is dear to him no more,
He answers prayers never, nor ever appears—
And so sore a loss is this to his lovers
They play never, the sweet reed sounds no more

In the oak coppice, or the Severn poplar shade
Silver-hearted . . . Softly wailing at eve
The silent country folk no more bring gifts
They delighted in—nor the new pipe greenly made.

The Poet

In nature's paid discernment
Quick and as steady still
As a bird or a beast of the grassment,
The poet learns his will,
Not by learning, not in obedience—but to fulfil.

And all he sees or hears
Will teach him, if he never
Lie in his reverence, for
Affection pays itself ever,
May go far on without chart or friend or cares.

If he know the song of the lark,
Rudely, as a brat calls it,
And the dawn, and water's teasing spark
Rushing in flood-time, and leaves' look
Week by week the year through it—
He will not harm book nor bark
But love, with lack, and golden truth, while devil spoils it.

Near Spring

Now the strong horse goes loose at last,
Free for his strength, in the February's end—
Floods have left the meadows and gone past
To the broader river's rocks and sand.

Musing on old tunes, ploughboys go
Hoping to catch the lilt forgot
Of a tune their fathers recalled in glow
Of talking, or after cricket hot.

The year stirs wing and watches skies
Deep-in again; the girl is happy
With her white apron showing in the doorway's
Frame; daffodils thrill in hedgebanks ruddy.

Where the Mire

Where the mire was thick from winter's sledges,
By the woodside, there I walked and mused
On the deep music floating there out of reach,
Just out of arm's length . . . It was December,
West country . . . Orange, low, the riband gleamed
Of the sunset, on it tangled twigs in teasing
Show, and the scent of the breath of the earth
Might have borne in me music to stir the silent
Folk of London; disowned, forgot of birth.
I thought that . . . and turning south to my end,
The farmhouse that too brought aid to my making,
Forgot much in tea's desire, by a hearth
To be sitting, talking countrywide, or till limned
Pictures in the fire glow—talking to world's ends . . .
In friendship hiding sorrow, in long thoughts of poetry
As gave us lines and reality of the apple orchard.

In December

In December the stubble nearly is
Most loved of things.
The rooks as in the dark trees are its friends
And make part of it . . .

Now when the hills shine far
And light and set off
That darkness, all my heart cries angrily
That music to fashion

For if not so, one must go
To the stubble every day
For comfort against such emptiness
As lost treasures make.

Cruelly scare the choughs from
Fallows and trees alike—
Though dim in love, or bright far
With the hills heroically they ally.

Here, If Forlorn

When to Mediterranean the birds' thoughts turn,
Watching the lessening days
And the softer glow
Of sunset, 'Goodbye' shall I say
And praise their beauty, and pray winter stern
To hurt nothing those feathers and fairy grace.
But after a week, in a place
Of coppices
I will count the kinds of birds that do not go
But for a Shakespeare and rare courage
Keep here, if forlorn
Despite sleeting scorn, and bitter hate of snow.

As They Draw to a Close*

As they draw to a close,
These songs of the earth and art, war's romanzas and stern . . .
The seaboard air encompasses me and draws my mind to sing
    nobly of ships . . .
Or the look of the April day draws anthems as of masters from
    me—
(O, it is not that I have been careless of the fashioned formal
    songs!)
For rough nature, for gracious reminder, I have sought all my
    days:
For men and women of the two-fold asking, for democracy and
    courtesy wherever it showed—
For the honour of flags well-borne at the heads of regiments
    digne . . .
Of what underlies my songs, the precedent songs, as they draw
    to a close I think,
Of failures rough crude half formless (yet I understood rarely
    why)
Of the blurred pictures of rare colour here and there shown on
    my pages . . .
Yet I have deserved well of men, and the book *Leaves of Grass*
    will show it—
Their homes and haunts nobler that I lived. (Hear the laugh
    ringing from the tavern . . .)
The meeting in market-place or hall, the workers together will
    remember me.
(In their talk are words like earth or panelled rooms, Baltimore,
    forced-march; page and maker-look—
The winds of the north still stir their eager questing minds.)

The seed I have sought to plant in them, trefoil, goldenrod and
    orchard-bloom,
These O precedent songs you also have helped plant everywhere
    in the world—
When you were launched there was small roughness in the touch
    of words,
A woman's weapon, a boy's chatter, a thing for barter and loss:

But I have roughed the soul American or Yankee at least to truth
    and instinct,
And compacted the loose-drifting faiths and questions of men in
    a few words.

*In this poem (and elsewhere), Gurney writes in the voice of Walt Whitman (author of
  Leaves of Grass).

Soft Rain

Soft rain beats upon my windows
Hardly harming.
But by the great gusts guessed farther off
Up by the bare moor and brambly headland
Heaven and earth make war.

That savage toss of the pine boughs past music
And the roar of the elms . . .
Here come, in the candle light, soft reminder
Of poetry's truth, while rain beats as softly here
As sleep, or shelter of farms.

OXFORD POETS

Fleur Adcock

James Berry

Edward Kamau Brathwaite

Joseph Brodsky

Michael Donaghy

D. J. Enright

Roy Fisher

David Gascoyne

David Harsent

Anthony Hecht

Zbigniew Herbert

Thomas Kinsella

Brad Leithauser

Derek Mahon

Medbh McGuckian

James Merrill

John Montague

Peter Porter

Craig Raine

Christopher Reid

Stephen Romer

Carole Satyamurti

Peter Scupham

Penelope Shuttle

Louis Simpson

Anne Stevenson

George Szirtes

Grete Tartler

Anthony Thwaite

Charles Tomlinson

Chris Wallace-Crabbe

Hugo Williams

also

Basil Bunting

W. H. Davies

Keith Douglas

Ivor Gurney

Edward Thomas